the materials of life

PATHWAYS IN SCIENCE

BIOLOGY 1—The Materials of Life
BIOLOGY 2—Built for Living
BIOLOGY 3—The Next Generation

EARTH SCIENCE 1—The Earth We Live On
EARTH SCIENCE 2—Oceans of Air and Water
EARTH SCIENCE 3—Man and Energy in Space

CHEMISTRY 1—The Materials of Nature
CHEMISTRY 2—Chemistry of Mixtures
CHEMISTRY 3—Chemistry of Metals

PHYSICS 1—The Forces of Nature
PHYSICS 2—Matter and Energy
PHYSICS 3—Sound and Light

GLOBE BOOK COMPANY, INC.

biology

1

the materials of life

JOSEPH M. OXENHORN
MICHAEL N. IDELSON

Edited by
PETER GREENLEAF

JOSEPH M. OXENHORN, now retired, has served as Head of Department of Biology and General Science, as Assistant Principal, as Elementary School Principal and Junior High School Principal in various New York City schools. Most recently, he was Principal of Theodore Roosevelt High School, Bronx, N. Y. He has headed science syllabus committees and has developed science films. Mr. Oxenhorn is author of Globe's *Pathways in Science* series, *Pathways in Biology* for high schools, and a college textbook of teaching methods for science teachers.

MICHAEL N. IDELSON has served as Chairman of the Department of Physical Sciences at Abraham Lincoln High School in Brooklyn, New York; Assistant Professor of Physics for the U.S. Naval Academy at Annapolis, Maryland, and President of the General Science Teachers Association of New York. He is also the author of three previous textbooks in the physical sciences.

PETER GREENLEAF has served as Chairman of the Science Departments at Montauk Junior High School and Eron Prep School and as an Instructor at the Cooper Union College of Engineering in New York. He was the Supervisor of Audio-Visual Instruction at the New York City Board of Education. He is the author of many science books and numerous related articles, films and filmstrips. He is now Adjunct Associate Professor in Science at Brooklyn College School of General Studies in New York City.

Ruth G. Oxenhorn served as editorial and research assistant in the original and revised editions.

Art by Ayres Houghtelling.
Text design by Nancy Dale.
Cover Photo: H. Armstrong Roberts

ISBN 0-87065-720-8 (soft cover)
ISBN 0-87065-735-6 (hard cover)

Editorial, art and production assistance provided by Center for Media Development, Inc.

PRINTED IN THE UNITED STATES OF AMERICA 0 9 8 7 6 5

Introduction to the Students

Welcome to your new science book, *Pathways in Science!*

A pathway is a road, a street or an avenue which leads you somewhere. *Pathways in Science* is a book that will lead you to a greater understanding of science.

This section of *Pathways* is called *The Materials of Life.* In this section, you will be studying a branch of science called "biology." Biology is the study of living things: what they look like, how they work and what they do. It is an exciting science and will answer many questions for you.

Pathways is a tool to help you learn. A good mechanic knows his tools. A good student knows how to use his book.

Let us study the organization of the book and see how to use it.

ARRANGEMENT

At the front of the book is a *Table of Contents.* This table or list tells you what you can find in the book. It tells you on what page each part of the book can be found.

The book is divided into four parts called *Units.*

Each Unit is divided into *Chapters.* Each chapter begins with a *Target Question* which is answered by material in that chapter. For example, on page 5, you will find Chapter 1. The Target Question is "How do we use our eyes to learn about plants and animals?"

Each chapter is divided into *Paragraphs* which have a number. Turn to page 5. There you will find paragraph 1, *Stop . . . Look . . . Learn.* Paragraph 2 on the same page is titled *Living Things Are Everywhere.*

Every few paragraphs, you will find a section called *You Now Know*. Turn to page 7 for an example. This is a kind of summary of what you just read. It helps you review. It helps you remember.

At the end of each chapter, you will find questions which test how much you have learned. Turn to page 10. There you will find the section called *Understanding What You Have Read*. In addition to the questions, you will also find suggestions for exciting things you can do at home. Look at page 10 again. Question IV is *For the Amateur Scientist*.

NEW SCIENCE WORDS

Science has its own language. It has new words which have a special meaning. Some of these new words are strange. The book tells you how to pronounce these words. Turn to page 5. In paragraph 2, we use the word BIOLOGIST. Notice how it is pronounced: (by-OL-o-jist).

At the end of each Unit, there is a *Glossary*. Turn to page 34. Notice that the words are arranged in alphabetical order. In Column 2, the words are written as they are pronounced. In Column 3, there are simple definitions.

PICTURES AND DIAGRAMS

The book has many pictures and diagrams to help you "see" science ideas. All pictures and diagrams are numbered in such a way that you can find them quickly. For example, 45–1 means the first picture on page 45; 69–2 means the second picture on page 69.

INVESTIGATIONS

Pathways has many investigations. Some of them you will see your teacher do in class. Some you will do in class by yourself or with a partner. Some you can do at home. Investigations are great fun and a good way to learn. The list of investigations appears on page ix.

SOME HINTS FOR A GOOD STUDENT

1. Listen carefully to your teacher.
2. Watch the investigations.

3. Get a good notebook. Keep it clean and neat.
4. Do your homework every day.
5. Read the book carefully. If you don't understand something the first time, read it again.
6. If you need more help, ask your teacher.
7. Don't mark up your book. Keep it clean.

AND REMEMBER . . .

School can be fun. School means learning. If you study today, you will succeed tomorrow. Good luck!

THE AUTHORS

List of Investigations

Table of Contents

UNIT I

Scientists are observers. Three great scientists, from top to bottom: Aristotle (Greek, 384–322 B.C.), Charles Darwin (English, 1809–1882) and Carolus Linnaeus (Swedish, 1707–1778). Life can be observed in the terrarium (middle) and aquarium (bottom). The magnifying glass (bottom) is a tool of observation and discovery.

IT'S FUN TO STUDY LIVING THINGS

UNIT I

It's Fun to Study Living Things

CHAPTER 1. WATCHING PLANTS AND ANIMALS AT WORK

CHAPTER 2. OBSERVING IS MORE THAN LOOKING

CHAPTER 3. WE STUDY LIVING THINGS IN THEIR NATURAL HOMES

CHAPTER 4. GROWING THINGS AT HOME

CHAPTER 5. WORKING IN A LABORATORY

1

TARGET
How do we use our eyes to learn about plants and animals?

Watching Plants and Animals at Work

1. STOP . . . LOOK . . . LEARN . . .

A familiar sign at railroad crossings reads: STOP! LOOK! LISTEN! This is almost the way scientists work, but they might change the sign to read: STOP! LOOK! LEARN! For much of what scientists know about living plants and animals was learned by doing just that. Someone sat very patiently and planned how he would find the facts he wanted. Then he looked very carefully to be sure he got his information right. And as a result, he learned a great deal.

2. LIVING THINGS ARE EVERYWHERE

Are you a Boy Scout or Girl Scout? Have you ever been to camp? If so, you have probably been on a "nature walk." You may have gone for a walk in the woods or to a pond or up a mountain.

Scientists may also go on nature walks to learn about living things.

A scientist who studies living things is called a BIOLOGIST (by-OL-o-jist). A NATURALIST (NAT-you-ruh-list) is a biologist who studies nature. A naturalist may walk in the woods, study wild animals in the jungle, or take a boat to a faraway island. Some have camped in the desert. Others have gone to the bottom of the ocean. A new way that scientists have found to gather facts has been to "walk" in space.

"But," you ask, "must I do this, too, if I want to learn about living things?" Of course not! You can watch the ants at the corner of a building. You can observe (watch) flies on a ceiling, birds on a telephone pole, and many animals in a zoo. Living things are everywhere. How can we study them?

3. LOOKING MORE CAREFULLY

About two thousand years ago, in far-off ancient Greece, there lived a wise man by the name of ARISTOTLE (AR-is-tot-ul). He was curious about many things. One of the things that interested him was how an egg develops into a chicken.

Like many other people, he had observed how the mother hen sat on the egg and warmed it until it hatched. Aristotle did more, however. He took an egg from a mother hen the day she began sitting on it, broke it open and looked at it very carefully. He then made a drawing of what he saw. He took a second egg from a mother hen after two days. Again, he broke it open, observed it carefully and drew a diagram. He did the same with a three-day egg, a four-day egg and so on for twenty-one days—which is how long it takes for an egg to hatch. Then Aristotle had the complete story. He was so careful in his work that in two thousand years only a few small errors have been found in his observations (ob-zer-VAY-shunz).

Fig. 6–1
Two thousand years ago, Aristotle studied how chickens' eggs hatch.

Fig. 6–2
An incubator keeps a steady temperature for hatching eggs. At what temperature is the incubator set? How do you know?

YOU NOW KNOW

- Seeing is observing.
- By observing, we learn about living things, their work and their needs.

4. OBSERVING HOW LIVING THINGS GROW

By observing carefully, Aristotle was able to learn how an egg grows to be a chicken. By observation *you* can learn how a chick sheds its young soft feathers (*down*) and grows to be a mature hen or rooster with bigger and tougher feathers. You can learn how babies grow to be adults, how a cut closes and forms a scar, how a flower bud opens into a full flower. In other words, by observation, you can learn about *growth*.

Fig. 7–1
The giant redwood at the left started life as tiny seeds. The giant cactus at the right grows in the desert. What are some ways to measure growth?

1. Some roses are red.
2. Violets are blue.
3. Daffodils are yellow.
4. Porcupines have spines.
5. Sheep have wool.
6. Lobsters have shells.

Fig. 8–1
Observation answers questions about appearance and structure. How might a blind person observe living things?

5. OBSERVING HOW LIVING THINGS ARE BUILT

If you sit quietly and watch a fly or a grasshopper, you will see that each of these animals has six legs. In the same way, you can discover that spiders have eight legs and that worms have none. The feathers of a parakeet's wings are soft, but the tail feathers are stiff. A fish has different kinds of fins, and not all its scales look alike. Using a mirror, look into your mouth and observe the different shapes of your teeth. Now look into your eyes. See how the color near the black spot in each eye is different from the color at the edges. By observation we can learn about the *parts* of plants and animals—their STRUCTURE (STRUK-cher).

6. OBSERVING HOW LIVING THINGS DO THEIR WORK

MAURICE MAETERLINCK (MAY-ter-link) was a famous writer and lawyer who became interested in bees. For over twenty years, Maeterlinck watched bees in their beehives. He saw how the queen bee was cared for and how the "worker" bees made honey. He kept careful notes and wrote a wonderful book called *The Life of the Bee.* If you have a cat or a hamster or a dog, you can observe your pet's habits, how it eats, sleeps, runs, fights, plays. You can watch an insect on a summer night fly around and around an electric light. You can see how a person breathes hard after exercise or a plant's leaves turn brown when it doesn't get water. In all these examples, you are observing how plants and animals do their work or, in other words, their BEHAVIOR (be-HAY-vyor).

Fig. 9–1

Careful examination of plants and animals is one important way to study biology.

YOU NOW KNOW

- Observation helps us learn about the STRUCTURE, GROWTH and BEHAVIOR of living things.
- Observation needs patience, time and care.

UNDERSTANDING WHAT YOU HAVE READ

I. What are the two main ideas of Chapter 1?

1. Observing plants and animals helps us understand how they carry on their life functions.
2. Aristotle was a great scientist.
3. Living things are everywhere.
4. Observation needs time and patience.

II. In which paragraph do you find the answer?

1. How does observing help us understand how young animals grow older?
2. How did a great scientist study bees?
3. What was one great discovery of Aristotle?
4. What is meant by behavior?

III. Knowing What and Why

1. Naturalists study
   a. nature.
   b. animals only.
   c. in a laboratory only.
   d. stars only.
2. Aristotle was the scientist who
   a. invented observation.
   b. studied the hatching of chickens' eggs.
   c. studied bees.
   d. learned about the structure of flowers.
3. A good example of the behavior of a bird is
   a. its size.
   b. its color.
   c. its feathers.
   d. its flying habits.
4. When a boy examines the parts of his dog's paws, he is studying
   a. structure.
   b. behavior.
   c. growth.
   d. life.

IV. For the Amateur Scientist

1. Collect leaves of some common trees in your neighborhood (*examples:* sycamore, maple, oak, linden, pine needles). Paste them on cardboard. Describe how each looks: color, shape, size, veins.

V. Go and Discover

1. *At a local park*
   Observe two different birds (example: pigeon and sparrow). List as many differences in their *activities* as you can discover. Now list as many differences in their *structure* as you can discover.
2. *At the local pet shop*
   Repeat the activity on *structure* above for two different kinds of fish (example: guppies and goldfish).
3. Repeat this activity, comparing cat and dog.

VI. Visit for Fun and Knowledge

Look in the Appendix, page 171. This lists zoos, aquaria and botanical gardens. Perhaps there is one in your area. Visits to these places will be fun and you will learn a great deal.

Suggestions:

1. Try to be present at animal feeding time.
2. Talk to the animal "keepers."

2

TARGET
What other senses help us study living things?

Observing Is More Than Looking

1. YOUR FINGERS CAN TELL YOU

So far we have used the word *observation* to mean only "seeing with our eyes." But, of course, we can also learn with our other senses. Have you ever grabbed a rose bush? If you have, your *sense of touch* probably told you about the sharp needles or *thorns*. Have you ever been stung by a jellyfish while you were swimming in the ocean? Your sense of touch probably told you to move away fast. It's easy to tell, even with your eyes closed, that a leaf is smoother than bark and that the fur of a collie is softer than that of a wire-haired terrier. Certainly you will never forget a porcupine once you have touched one!

Fig. 11–1
Which sense will teach you the difference between these two animals?

2. YOUR NOSE CAN TELL YOU

You can also use your sense of smell to learn about living things. Flowers are beautiful to look at. Many also have a pleasant odor that helps you recognize them. You can tell when peppermint is growing in a garden by its sharp, pleasant smell. Pine woods are easy to recognize even if your eyes are closed. Books about plants often describe them by their *odor*. The smell of a skunk is not easy to forget!

3. TASTING TO LEARN

Scientists also identify living things by taste—another one of our five senses. We say that apples taste sweet or sour. We speak of sweet strawberries, salty fish, sour grapes and bitter almonds. Wild animals learn quickly not to bite into certain plants because of their bitter, sour or sharp tastes.

Which jar contains salty water?
Which jar contains sweet water?

Fig. 12–1
Your tongue can answer the question.

4. LISTEN AND LEARN

Finally, we also use our hearing to learn about living things. A dog's bark, a cat's meow and a bird's song are probably very familiar to you. In the quiet of the country, one can hear the crickets chirping, the locusts "singing" and the robins warbling. By listening to your heart, the doctor can tell you about your health.

Fig. 12–2
Different kinds of birds have different songs. By listening carefully, we can tell one kind of bird from another.

YOU NOW KNOW

- We can study living things by observation with our five senses: seeing, hearing, touching, tasting and smelling.

5. SCIENTISTS KEEP RECORDS

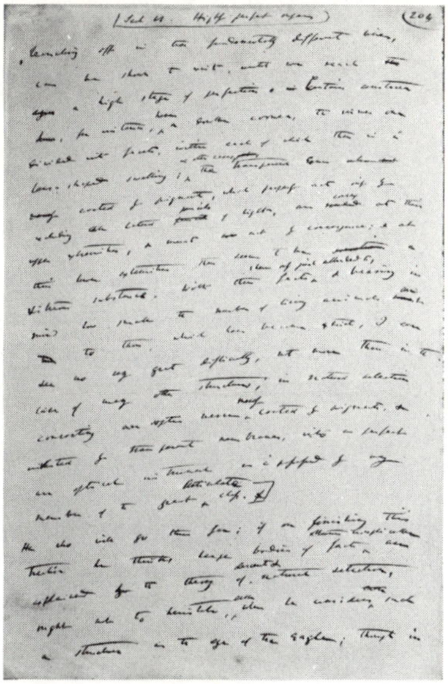

Fig. 13–1

A page from the notebook of the scientist Charles Darwin. Why is a notebook important to a scientist?

Scientists find it useful to keep records of what they observe by writing in a notebook. Quite often they also make drawings of what they see. What is the value of this? Notes help them remember things which they might otherwise forget. Secondly, scientists can compare their own observations with those of other scientists. In addition to diagrams, modern scientists also use cameras and slow-motion photography as a way of keeping records. A motion picture, slowed down, might show how a bird flies. Speeded up, it might show how a plant grows from bud to blossom. Scientists now also use the tape recorder to catch the sounds of birds, insects and other animals.

YOU NOW KNOW

- Scientists keep notes, draw diagrams, take photographs and make tape recordings.

6. STUDENT SCIENTISTS ALSO KEEP RECORDS

As a student in science, you will certainly want to keep records of what you learn. Do you have a good notebook? Do you write neatly? Do you underline properly? Do you use a ruler? Do you enter the date on each piece of work? If you can draw, why not add a sketch of what you see? If you have a camera, you may want to take a picture. Or you can search the magazines and newspapers for pictures to paste into your notebook. At the end of the chapter, you will find several home experiments which require that you keep records. Try them.

Fig. 13–2

A good student keeps a good notebook.

UNDERSTANDING WHAT YOU HAVE READ

I. What are the two main ideas of Chapter 2?

1. Common sense helps you learn.
2. We learn with all five senses.
3. Keeping records of our observations helps us understand better.
4. All good scientists use cameras and tape recorders.

II. Explanation please . . . answer in one or two sentences.

1. How is touch a way of learning?
2. Why do scientists keep notebooks?
3. How many senses do we have?
4. How is a good student like a good scientist?

III. Knowing What and Why

1. Scientists keep notes of their observations because
   a. they can review their observations later.
   b. it is a custom in science.
   c. notes explain diagrams.
   d. they need notes to prove they saw these things.
2. Tasting a strange-looking leaf in the woods is not always a good way to study the plant because
   a. tastes differ.
   b. taste tells you nothing.
   c. tastes change according to the season.
   d. some leaves can make you sick.
3. Slow-motion cameras would be *least* useful in studying
   a. a tree.
   b. a bird.
   c. a lobster.
   d. an ant.
4. Now that we have cameras, scientists
   a. never draw diagrams anymore.
   b. draw diagrams of plants only.
   c. draw diagrams and take photographs.
   d. draw diagrams only in laboratories.

IV. Using Your Judgment

1. Suggest how *each* of the following might be used to study the animal named.
   a. a tape recorder for crickets
   b. a motion-picture camera for bees
   c. a still camera for an eagle
   d. your fingers for birds
   e. a watch for frogs
2. Explain why the following statements are correct:
   a. To understand how a beaver builds a dam, it is a good idea to observe the work several days in a row.
   b. Strawberries can be judged by their color, taste and weight.

V. For the Amateur Scientist

1. Collect pictures of plants which are used to improve the taste of foods: mustard, cloves, pepper, peppermint.
2. Collect pictures of different kinds of dogs. Get information about their fur, weights and heights.

3

TARGET
Where does the scientist study living things?

We Study Living Things in Their Natural Homes

1. BIOLOGISTS ARE EXPLORERS

As we have already seen, we can learn a great deal of biology without traveling far from home. But biologists sometimes have to be explorers in far-off places.

Let us take as an example the English biologist CHARLES DARWIN. In 1831 he sailed on a ship called the

Fig. 15–1
You too, can be an explorer without traveling far from your home.

A pair of field glasses or
BINOCULARS (by-NOK-yew-lerz)
for studying birds

A net for
collecting
insects

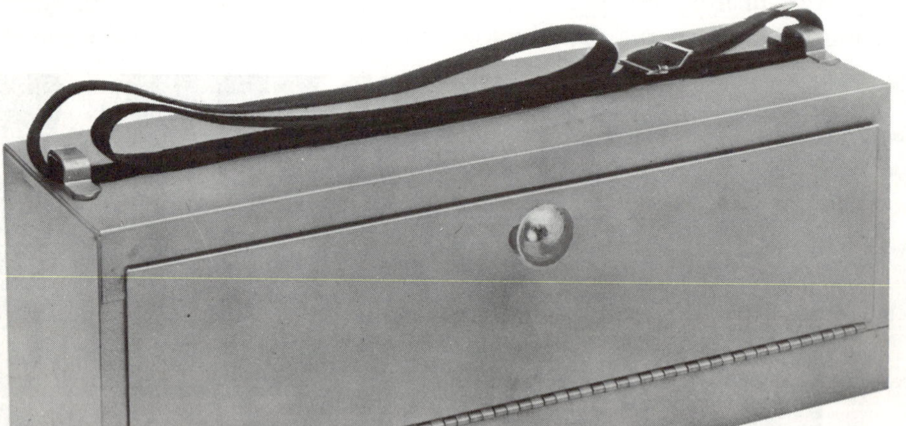

A VASCULUM (VAS-kew-lum)
for collecting fresh plants

Fig. 16–1

Tools of the naturalist. Give uses for each piece of equipment pictured. Which of these can be homemade?

Beagle and visited faraway islands and the shores of South America. He did not return until 1836—five years later. He brought back with him many notebooks filled with observations about birds, insects, coral and fish. He also brought back a large nature collection. Later, you will learn about Darwin's important contribution to science.

Another famous expedition was that of CAROLUS LINNAEUS (lin-NAY-us), a Swedish naturalist. In 1729 Linnaeus began an exploration of Lapland in the Far North. He made most of this journey on foot, making observations, collecting samples and making drawings. As a result of his explorations, he developed a system for naming plants and animals. Here again, the scientist turned explorer. Today, with modern transportation, scientists travel far and wide to observe, to collect and to learn.

2. A WALK IN THE WOODS

A walk in the woods will give you a chance to see many living things. You will see tall pine trees, trailing berry bushes and scampering chipmunks. You will see a squirrel stop to nibble on a nut. You will catch the flash of red on a bird's wing. Turn over a log and you will see bugs hurry out of sight. You may be frightened by the rapid movement of a garter snake or the whine of a green fly. A beautiful giant butterfly may stop one inch from you and disappear in a second when you try to catch it.

Naturalists often collect samples to bring home to study. A nature collection of insects and leaves can be the beginning of a wonderful hobby.

3. AT THE SEASHORE

Walk along the ocean front some day. You will see all kinds of living things that come from the sea. The beach probably has many shells that once held living oysters and clams. If you dig below the wet sand, you may be lucky enough to find some live shelled creatures—a mussel or clam or a little sea crab. You can learn to identify shells very quickly.

4. ON THE OCEAN AND UNDER THE OCEAN'S SURFACE

Fishing is for fun and food. But fishing is also for study. Many scientists have spent hours on boats, catching fish to study and dissect. Perhaps you yourself have gone crabbing. When you raise the crab basket, you have a chance to study the structure and behavior of crabs. Have you ever traveled in Maine or in Canada? Then you have probably seen lobster traps. Raising a lobster is a promise of a good meal. But scientists also learn about lobsters and their habits by catching them for observation. Studying fish and other kinds of animals under the ocean has become very easy with modern diving equipment. One of the most fascinating adventures has been in a BATHYSCAPHE (BATH-uh-skafe), a kind of submarine that can travel to the ocean floor. From inside a bathyscaphe scientists can use special cameras to take pictures of ocean creatures.

Fig. 18–1
William Beebe developed this bathysphere.

Fig. 18–2
Navy deep-sea divers bring back information about life under the ocean's surface.

5. A TRIP IN THE DESERT

The word *desert* probably brings to your mind a great deal of sand, very little water and not much life. But biologists have explored deserts and have found living things of great interest, such as desert plants, cactus and lizards. There are many scientists who go to deserts on "digs." By excavating (digging) in sand, they sometimes find bones and fossils of animals that lived very long ago.

Fig. 18–3
At the wheel: Dr. Roy Chapman Andrews—scientist and explorer in the Gobi Desert in China.

Fig. 19–1

ASTRONAUTS (AS-tro-nawts) brought back information about how human beings react to space.

6. A "WALK" IN SPACE

If you have ever read any science-fiction stories, you might think that space travel is an easy undertaking. Actually, it presents many problems. One of the most important problems has been whether or not man can exist in outer space. Of course, we now know that he can. We have found this out in several ways. First, we tested spacemen on earth under conditions like those in space. Second, we sent animals such as mice and dogs into space to see how well they would survive. Finally, we sent men into space, giving them many instruments to take along. These instruments tell us how well the men survive, by checking their rate of breathing, their heartbeat and their blood pressure. American and Russian spacemen have "walked" in space. The instruments they carried with them told us much about conditions in space and man's ability to travel safely in space. This information has also given us ideas for future spaceships and protective space suits.

7. THE HOME OF LIVING THINGS

All living things share the earth. They live in a shallow layer all around the earth. This layer includes the land, the water, and the air over land and water. This shallow layer where animals and plants can live is called the BIOSPHERE (BY-oh-sfeer). It is the home of life on earth.

The study of life in the biosphere is called ECOLOGY (ee-KOL-oh-jee). In the study of ecology, the scientist learns about the *relationships* of one kind of living thing to all other living things. He also studies how the living things are affected by the *nonliving* parts of the biosphere. Fig. 20–1 shows these nonliving factors.

8. LIFE REGIONS ON EARTH

There are many different kinds of environments in the biosphere. They have different conditions. Therefore, different kinds of plants and animals are found in different kinds of "homes." Each different kind of home is called a BIOME (BY-ohm). For example, cactuses and lizards are found in the *desert*. Visit the *grassland*, and you will see grasses, foxes and buffalo. The beautiful orchids and the smart little monkeys make their home in the *tropical rain forest*. Altogether, there are six land biomes. There are two kinds of water biomes: the fresh water biome of lakes and rivers, and the marine (salt water) biome of bays, seas, and oceans. As you can guess, the plants and animals in these two water biomes are different from each other. Fig. 20–2 shows the eight biomes of the earth.

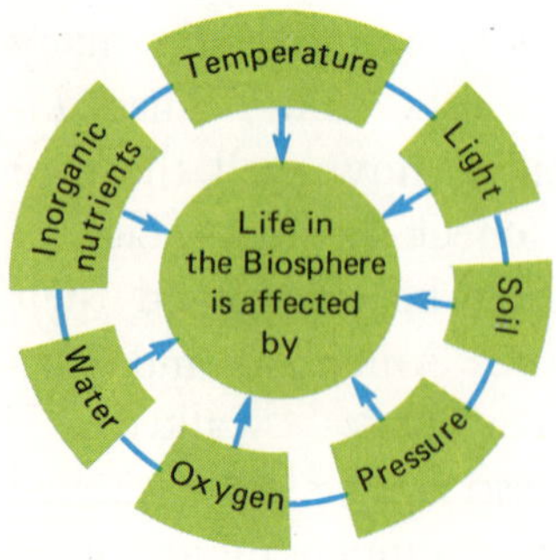

Fig. 20–1
Nonliving factors in the biosphere affect plants and animals. How?

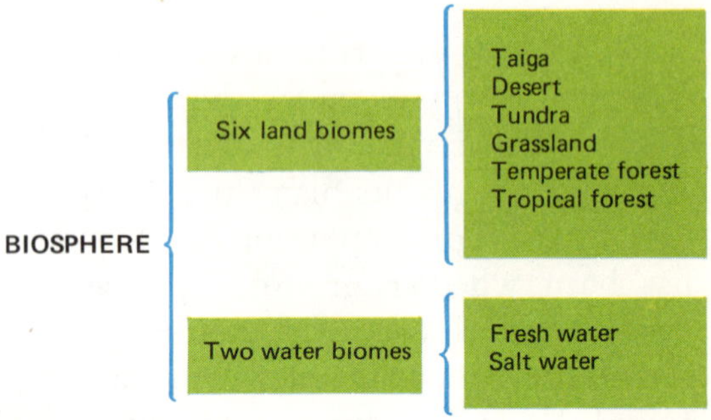

Fig. 20–2
The biomes of the earth have different conditions. Look up the kinds of life found in each kind of biome.

UNDERSTANDING WHAT YOU HAVE READ

I. What's the Difference?

State the difference between each of the following pairs of words:

1. biology-ecology
2. biosphere-atmosphere
3. marine water-fresh water
4. naturalist-astronaut

II. Explanation please . . . answer in one or two sentences.

1. Where do naturalists go to collect samples?
2. Why do scientists study the effects of space travel on human beings?
3. What do excavations sometimes uncover?
4. How do scientists study living creatures below the ocean surface?

III. Knowing What and Why

1. Biologists sometimes go to distant places to study living things because
   a. it's hard to study nature close to home.
   b. they can then see living things in their natural homes.
   c. they can meet other scientists there.
   d. they can try out new equipment.
2. Naturalists collect plants and animals because
   a. this is their hobby.
   b. they need samples.
   c. they can study them at a later time.
   d. they want to interfere with living things.
3. Which piece of equipment would you *not* use on a nature walk through the woods?
   a. binoculars
   b. butterfly net
   c. vasculum
   d. bathyscaphe
4. When biologists study living things, they
   a. must travel to far-off places.
   b. must collect samples.
   c. must take photographs.
   d. must keep careful records.

IV. Match the Job with the Scientist

1. William Beebe
2. Charles Darwin
3. Carolus Linnaeus
4. Roy Andrews

a. a visit to Lapland
b. exploring faraway islands
c. exploring the ocean
d. exploring the Chinese desert
e. sailing the *Kon-Tiki*

V. For the Amateur Scientist

1. Search old magazines. Collect pictures of faraway places which you would like to visit someday to study living things. Bring your collection to class.
2. Visit your local museum, botanic garden or zoo. Take some pictures of plants and animals. Bring your pictures to class.
3. Join the Boy Scouts, Girl Scouts or 4-H Club in your community. Join their nature group.

4

TARGET
How can we grow things at home for observing and experimenting?

Growing Things at Home

1. LIVING THINGS AT HOME

People have many different kinds of pets: dogs, cats, hamsters, canaries and parakeets. Many homes have a few plants on the window sill and perhaps a fish tank. Living things are interesting. They make a home pretty and colorful. As a young scientist, you may be interested to know that you can grow many new things at home. This will make it possible for you to observe and experiment for yourself.

2. LIFE IN AN AQUARIUM

A fish tank is also called an AQUARIUM (uh-KWAIR-ee-um). An aquarium contains fish and green water plants. You can see in an aquarium how the fish swim, how they take in food and how they take in water. Green plants take in carbon dioxide and give off oxygen. Fish use oxygen and give off carbon dioxide. Later, you will learn how fish remove oxygen from the air in the water. When an aquarium has the right number of fish and plants, the two live well together. The water has to be changed only once in a great while. We say then that the aquarium is *balanced*.

3. LIFE IN A TERRARIUM

A TERRARIUM (te-RARE-ee-um) is a glass tank or bowl filled with soil, not water. In a good terrarium the soil is dark and rich. Moss plants are grown on top of the soil. Ferns grow above the soil. A terrarium is covered with glass because this keeps the atmosphere inside very moist. A few pieces of bark and a rock or two make a very attractive scene, and a frog or lizard is sometimes included to complete the "family."

Fig. 23–1

The aquarium and terrarium are small "worlds" of living things. How can life go on inside these containers even when they are closed?

4. GROWING PLANTS FROM SEEDS

The seeds that some plants produce are alive and can grow into plants if they have water and proper temperature. When seeds begin growing roots

DO AND DISCOVER

Investigation 1: How can we observe germinating seeds?

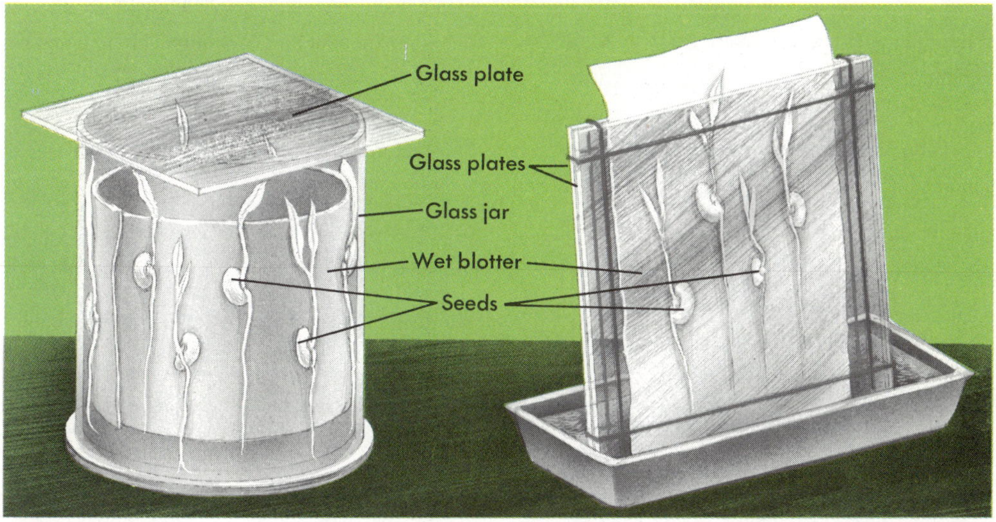

Line a jar with a wet blotter. Place seeds between the blotter and the sides of the jar. Cover the jar with a saucer or glass plate.

Place a blotter between two pieces of glass. Place the seeds on the blotter and use rubber bands to hold the plates together. Stand the plates in a tray of water.

Fig. 23–2

Two easy ways to germinate seeds at home. Try them. Could these same set-ups be used to check the way temperature affects germination? Explain.

DO AND DISCOVER

Investigation 2: How can we grow plants from cut stems?

Fill a jar with water. Keep refilling it as the water evaporates.

Fill a coffee can with sawdust. Place a cutting in the soil inside the flower pot. Keep the soil and sawdust wet.

Fig. 24–1

The cutting shown above has three nodes. *Nodes* are the thicker parts of stems.

and stems, we say they GERMINATE (JER-min-ate). Most seeds, if soaked in water overnight, will germinate faster. Corn seeds, black beans and pea seeds germinate fast and show roots and stems clearly.

5. PLANTS FROM CUT STEMS

Many plants when cut through the stem and placed in water, soil or sand will grow roots. Plants which are cut at the stems and then used to grow new plants are called *cuttings*. Common house plants such as GERANIUM (juh-RAY-nee-um), BEGONIA (buh-GO-nee-ya) and COLEUS (KO-lee-us) can be cut and regrown in this way. Each cutting will become a new plant.

6. WE GROW PLANTS FROM PARTS OF PLANTS

Do you know which part of a plant you eat for food? Here are four common plants:

Onion—a stem, with leaves wrapped around it
Sweet Potato—a fat, sweet root
White Potato—an underground stem
Carrot—a root

DO AND DISCOVER

Investigation 3: How do common plants multiply?

Fig. 25–1

What three conditions will be needed to make each plant grow? Try to invent an investigation to check each one.

The pictures above show how parts of a plant can be used to grow a whole plant.

7. INVISIBLE LIVING THINGS

The air, soil and water are filled with living things which we cannot always see. These are plants that must be studied with a MICROSCOPE (MY-kro-skope). Scientists have discovered that the air is filled with tiny SPORES (SPORZ) which, like seeds, can germinate to form tiny plants called MOLDS. Perhaps you have seen bread mold or molds on cheese or oranges. Molds spoil valuable food. For this reason, food must be kept covered.

8. TRAPPING AND GROWING MOLD SPORES

Obtain a shallow glass baking pan. Line its bottom with blotting paper which you have soaked in water. Lay

DO AND DISCOVER

Investigation 4: How can we trap and grow mold spores?

out the foods shown above. Leave the pan uncovered overnight. If you are lucky, some spores may fall on the foods. The next day cover the pan with a piece of Saran Wrap. Then put it away in a dark, warm place for a few days.

DO AND DISCOVER

Investigation 5: How can we grow earthworms?

Fill half of a large jar with a mixture of soil, decayed leaves, grass, lettuce and coffee grounds. After a heavy rain, you can usually find some worms on the surface of a garden or near a bush. Drop two or three worms into the mixture. Keep it moist but not too wet. Keep the jar covered with a screen or a piece of cheese cloth.

9. EARTHWORM AND ANT COLONIES

Some people may feel a little uncomfortable about having earthworms and ants in their homes. But if properly cared for, they are interesting and harmless.

DO AND DISCOVER

Investigation 6: How can we develop ant colonies?

Your "mix" here is moist, sandy soil. Drop a small ant hill into the jar. Wrap brown paper around the jar. For food, add tiny bits of lettuce, carrot and scraps of meat. Remove food which is not used. When you want to observe the ants, pull up the brown wrapper.

UNDERSTANDING WHAT YOU HAVE READ

I. What are the two main ideas of Chapter 4?

1. An aquarium contains fish.
2. We can learn about living things by growing them at home.
3. Molds spoil bread.
4. It is easy to grow living things at home.

II. Explanation please . . . answer in one or two sentences.

1. How can we keep an aquarium "balanced"?
2. What part of the onion plant do we eat?
3. How can we study the cause of moldy bread?
4. How can we grow many plants from one plant?

III. Knowing What and Why

1. You may have to change the water in an aquarium very often, if
   a. mold spores fall in.
   b. there are too many fish.
   c. you don't feed the fish.
   d. there are no plants in the tank.
2. The following are *not* wanted in a terrarium, except
   a. ants.
   b. molds.
   c. fish.
   d. lizards.
3. Which of the following are needed for seeds to germinate?
   a. water and warmth
   b. water and darkness
   c. water and light
   d. darkness and warmth
4. When we use cuttings to grow new plants, which parts are usually missing?
   a. stems and leaves
   b. stems and roots
   c. roots only
   d. roots and leaves

IV. Fill in the Blanks

1. New molds can grow from ______.
2. New geraniums can grow from ______.
3. A carrot is the ______ of the plant.
4. In a terrarium, there are no ______.
5. In an aquarium, there are no ______.

a. cuttings
b. hamsters
c. spores
d. bacteria
e. ants
f. root
g. leaves

V. For the Amateur Scientist

1. Visit a local nursery or greenhouse where plants are sold. Ask the owner to show you his cutting benches.
2. Visit a tropical fish store. Explain to the class how it is possible for many fish to live in a tank that has no plants at all.

VI. Earning Extra Money

Buy five or six very healthy geranium plants and five or six very healthy begonia plants. Make cuttings and plant them in pots. If you do this indoors in February, you may be able to sell your plants, house to house, for Easter. You can make a handsome profit!

5

TARGET
How are plants and animals studied in a laboratory?

Working in a Laboratory

1. A LABORATORY IS MORE THAN A WORKSHOP

You may have heard people call a laboratory a *workshop*. But how is a laboratory more than just a workshop? In a laboratory we EXPERIMENT (ex-PER-i-ment). This means that we try to solve problems and answer questions about science. This can be done in many ways. Sometimes scientists try out certain answers or solutions and find that they are wrong. So they try again. By trying out many wrong answers and solutions, scientists sometimes finally find the right one. We call this method of experimenting *trial and error*.

Let us look at some of the instruments used in the laboratory.

Fig. 28–1
This laboratory is a scientist's workshop. Are all laboratory workers college-trained scientists? Explain.

2. USING INSTRUMENTS IN THE LABORATORY

In order to carry out experiments, scientists need many kinds of APPARATUS (ap-uh-RAT-us), or instruments. Some instruments help scientists get exact answers by measuring or weighing the results of their experiments. Some instruments are used to control the growth of living things, so that living things can be better understood. Where do these instruments come from? There are hundreds of factories in the United States employing thousands of workers whose business it is to make instruments used by scientists. Some day, you may be able to get a job in such a plant.

3. HOW BIG? HOW HEAVY? HOW LONG?

How wide is the small intestine? How high is an oak tree? How long is an earthworm? These are simple questions you can answer by measuring with a ruler, a yardstick or a tape. Weighing living things or their parts is another form of investigating. For example: Which weighs more—a cat's brain or a monkey's brain? Does a diseased kidney weigh more or less than a healthy kidney? The scale, or CHEMICAL BALANCE, provides the answers. A watch tells how long it takes to digest a meal or helps to count the number of breaths a person takes in a minute.

Fig. 29–1
Laboratory scales can be used to measure very small differences in weights. What units of weights are usually used by scientists?

YOU NOW KNOW

- Simple measuring instruments are used to measure size, weight and time.

4. HOW HOT? HOW COLD?

All living things react to great changes in temperature. It should not surprise you, then, to know that a THERMOMETER (ther-MOM-uh-ter) is one of the most important laboratory instruments. It answers the questions "How hot is it?" and "How cold is it?" Or, more exactly, "What is the temperature?" In studying living things, we often wish to see how they behave or respond when the temperature changes. Or we may actually wish to measure the body temperature of an animal.

5. INSTRUMENTS FOR THE DOCTOR

When we think of a plumber, we usually think of a wrench. When we think of a doctor we almost always think of his STETHOSCOPE (STETH-o-skope). This wonderful instrument helps your doctor listen to the way your heart beats so he can judge your health. We might also think of his X-ray camera. This instrument can photograph the bones in your body. X-rays pass through skin and flesh so that only bones show in the picture. In this way the doctor can tell whether you have a broken bone.

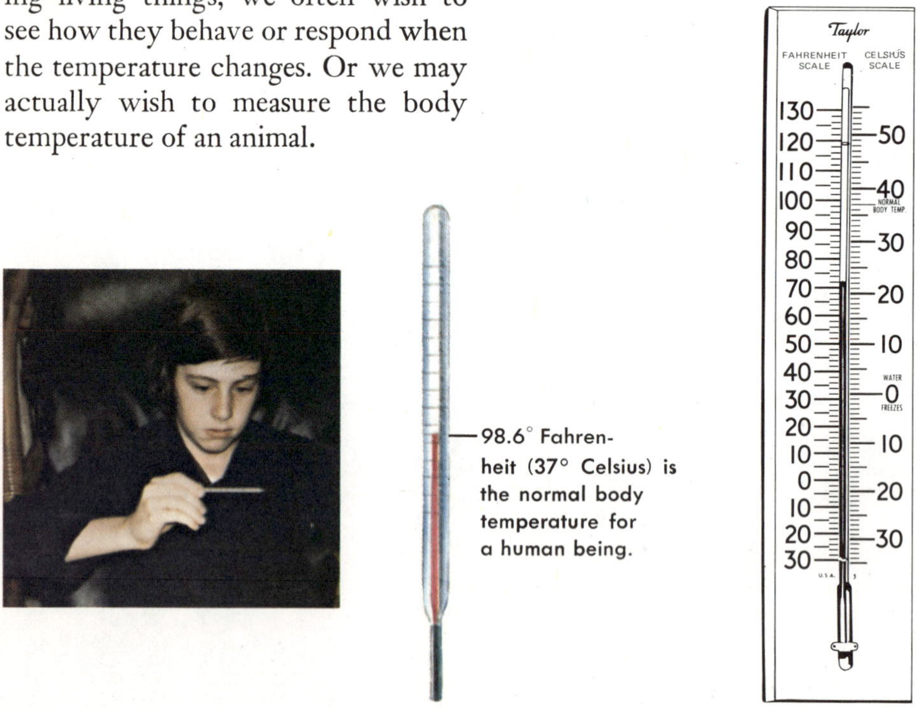

Fig. 30–1
The thermometer is used to check body temperature. Can you tell what 98.6° F. equals on the Celsius scale?

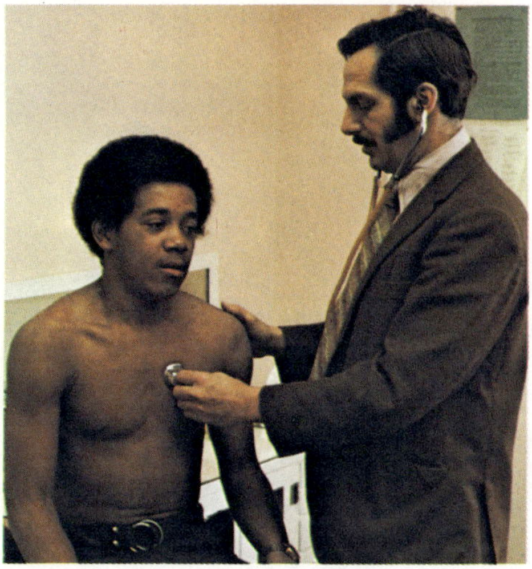

Fig. 31–1
The stethoscope helps the doctor listen to your heartbeat.

6. CHEMICAL TESTS SHOW THE WAY

Perhaps you know that your exhaled breath causes clear limewater to turn cloudy. You may recall that this shows that our exhaled air contains carbon dioxide. It has been known for a long time that living things can react with chemicals and will show certain changes. This is of great help when we study plants and animals. There is a chemical test called URINALYSIS (YUR-in-AL-i-sis). URINE (YUR-in) is the liquid waste of animals. If the analysis shows traces of sugar in the urine, this indicates that the person has a disease called DIABETES (dy-a-BEE-teez).

Ask your teacher for a strip of red litmus paper. Place a drop of saliva on the paper. It will probably turn light blue. This is a test for the chemical nature of saliva. Ask your teacher for a strip of PTC paper. Taste the paper. PTC is a chemical that tastes salty, sweet, sour or bitter to different people. Most people will not taste PTC at all. The chemical PTC is used to study the taste power of human beings.

YOU NOW KNOW

- Chemical tests can be used to study behavior and life functions.

7. A LOOK INSIDE LIVING THINGS

Have you ever taken a toy apart? Or an old clock? You were really playing the part of a scientist. You were curious and you wanted to know "how it worked." Scientists also want to look inside plants and animals to see "how they work." They do careful cutting called DISSECTION (di-SEK-shun).

When scientists dissect plants, they can study the insides of roots and stems. In animals, they can see how muscles are attached to bones, how many sections a brain has, the parts of the heart, and many other things. Of course, when scientists dissect, they are careful not to cause pain to an animal. They use painkillers or put the animal to sleep in a painless way. A true scientist never does cruel things to animals.

YOU NOW KNOW

- Dissection is the careful cutting of parts of plants and animals to see how inside parts look and work.

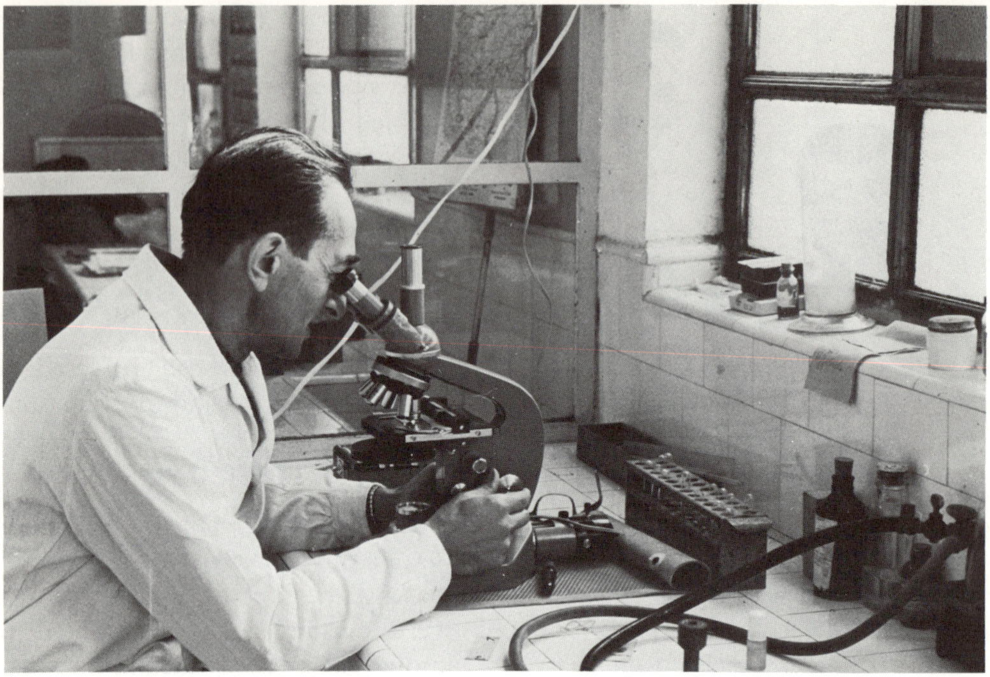

Fig. 32–1

THE MARK OF A REAL SCIENTIST: he never performs cruel experiments on any living thing.

UNDERSTANDING WHAT YOU HAVE READ

I. What are the two main ideas of Chapter 5?

1. Laboratories have expensive equipment.
2. There are many instruments that help us study science.
3. The most important instrument in a doctor's office is the stethoscope.
4. Dissection helps us look inside animals.
5. Experiments in the laboratory make use of many instruments.

II. In which paragraph do you find the answer?

1. How is temperature measured?
2. Is taste power different in different people?
3. What is dissection?
4. What instrument helps us see bones through flesh?

III. Knowing What and Why

1. Instruments such as thermometers, scales and rulers are used by scientists because
   a. they are accurate and take away guesswork.
   b. they are always right.
   c. they give us information.
   d. they are very easy to use.
2. The scientist who makes a mistake
   a. should drop the experiment.
   b. should use instruments.
   c. should try to find the error.

d. should look up the information he wants in a good book.

3. In using chemicals to test the activities of plants and animals, it is most important
   a. to use the right amount of the chemicals.
   b. to be sure the chemicals are not poison.
   c. to know that the chemicals will always give the same results.
   d. to be sure the color, odor and taste are right.
4. Even though X-rays show you the bones inside an animal, dissection is still important because
   a. X-rays are too costly.
   b. X-rays do not show flesh and skin.
   c. X-rays are not good for plants.
   d. dissection can be done by anyone.

IV. Fill in the Blanks

1. Another word for apparatus is ______.
2. Thermometers measure ______.
3. Trial and error is a method of finding the ______.
4. PTC tests ______.
5. A doctor uses a stethoscope to check your ______.

a. barometer
b. correct answer
c. taste
d. temperature
e. instruments
f. heartbeat

V. On the Ladder of Understanding

1. Radioactive iodine is an ______ of ordinary iodine.
2. Carbon dioxide causes cloudiness in the chemical called ______.
3. Carbon dioxide is a compound of the elements ______ and ______.

VI. For the Amateur Scientist

1. Using a watch with a second hand, count how many times you breathe in and out in one minute. Do some vigorous exercise. Check your breathing rate again. Report to your class on the difference noted.
2. Which is longer, line AB or line CD?

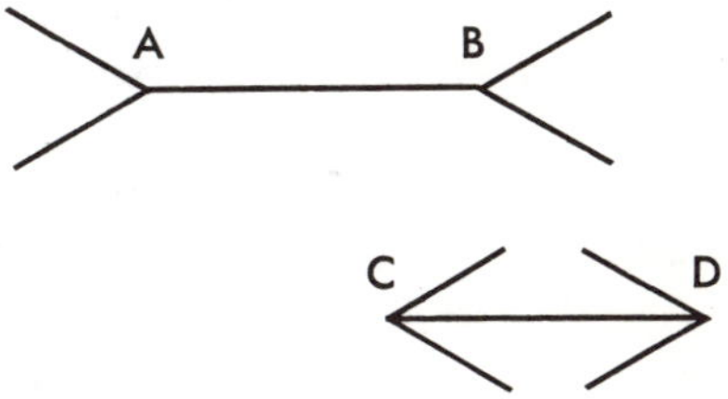

Now measure each line with a ruler. Explain why this proves that scientists need instruments.

3. Pick up your baby brother or baby sister. Can you guess his or her weight? Now weigh the child on a scale. How close was your guess to the actual weight?

Glossary: What Does It Mean?

Andrews, Roy		Explorer who found fossils of dinosaurs in Asia.
apparatus	ap uh RAT us	Instruments and equipment.
aquarium	uh KWAIR ee um	A fish tank.
Aristotle	AR is tot ul	Greek scientist who studied the hatching of chickens' eggs two thousand years ago.
astronauts	AS tro nawts	People who travel in space.
bathyscaphe	BATH uh skafe	A type of research submarine that can dive to the ocean's floor.
begonia	buh GO nee ya	A common house plant.
behavior	be HAY vyor	The way in which plants and animals act or do their work.
binoculars	by NOK yew lerz	Field glasses which enlarge objects at a distance.
biologist	by OL o jist	A scientist who studies living things.
biome	BY ohm	One of the eight regions of the earth where living things are found.
biosphere	BY oh sfeer	The spherical zone surrounding the earth which forms the home of living things.
chemical balance		A form of laboratory scale.
coleus	KO lee us	A common house plant.
Darwin, Charles	DAR win	English biologist.
diabetes	dy a BEE teez	A disorder in which sugar is not burned up by the body.
dissection	di SEK shun	Careful cutting of parts of plants and animals to see how their insides look and work.
down		Soft feathers of a young bird.

ecology	ee KOL oh jee	The study of relationships among living things and their response to the nonliving environment.
experiment	ex PER i ment	A controlled study in a laboratory to find an answer to a problem.
geranium	juh RAY nee um	A common house plant.
germinate	JER min ate	To grow from a seed into a plant.
incubator	IN kew bay ter	A machine in which the temperature is controlled—used for hatching eggs.
Kon-Tiki	kon TEE kee	A raft which sailed across the Pacific Ocean.
Linnaeus, Carolus	lin NAY us	A Swedish naturalist who developed scientific names for plants and animals.
Maeterlink, Maurice	MAY ter link	Belgian author who wrote *The Life of the Bee.*
molds		Small plants which grow on food.
naturalist	NAT you ruh list	A scientist who studies nature.
nodes	NOHDS	The swellings on the stem of a plant.
observation	ob zer VAY shun	Act of watching closely.
spores	SPORZ	Parts of a mold used to reproduce new molds.
stethoscope	STETH o skope	A doctor's instrument for listening to the heartbeat.
structure	STRUK cher	The way a thing is built or put together.
terrarium	te RARE ee um	A tank filled with soil and plants.
thermometer	ther MOM uh ter	Instrument for measuring temperature.
thorns		Sharp needles on some plants such as rosebushes.
trial and error		Method of trying out different ways to solve a problem.
urinalysis	YUR in AL i sis	A chemical test of urine.
urine	YUR in	Liquid waste of animals.
vasculum	VAS kew lum	A metal basket used on field trips to collect plants.

UNIT II

The life functions of living things include taking in foods, digesting foods, and excreting wastes. Life is found in the air, on land, and in water. The microscope (bottom left) opens a new window into life.

WHEN THINGS ARE ALIVE

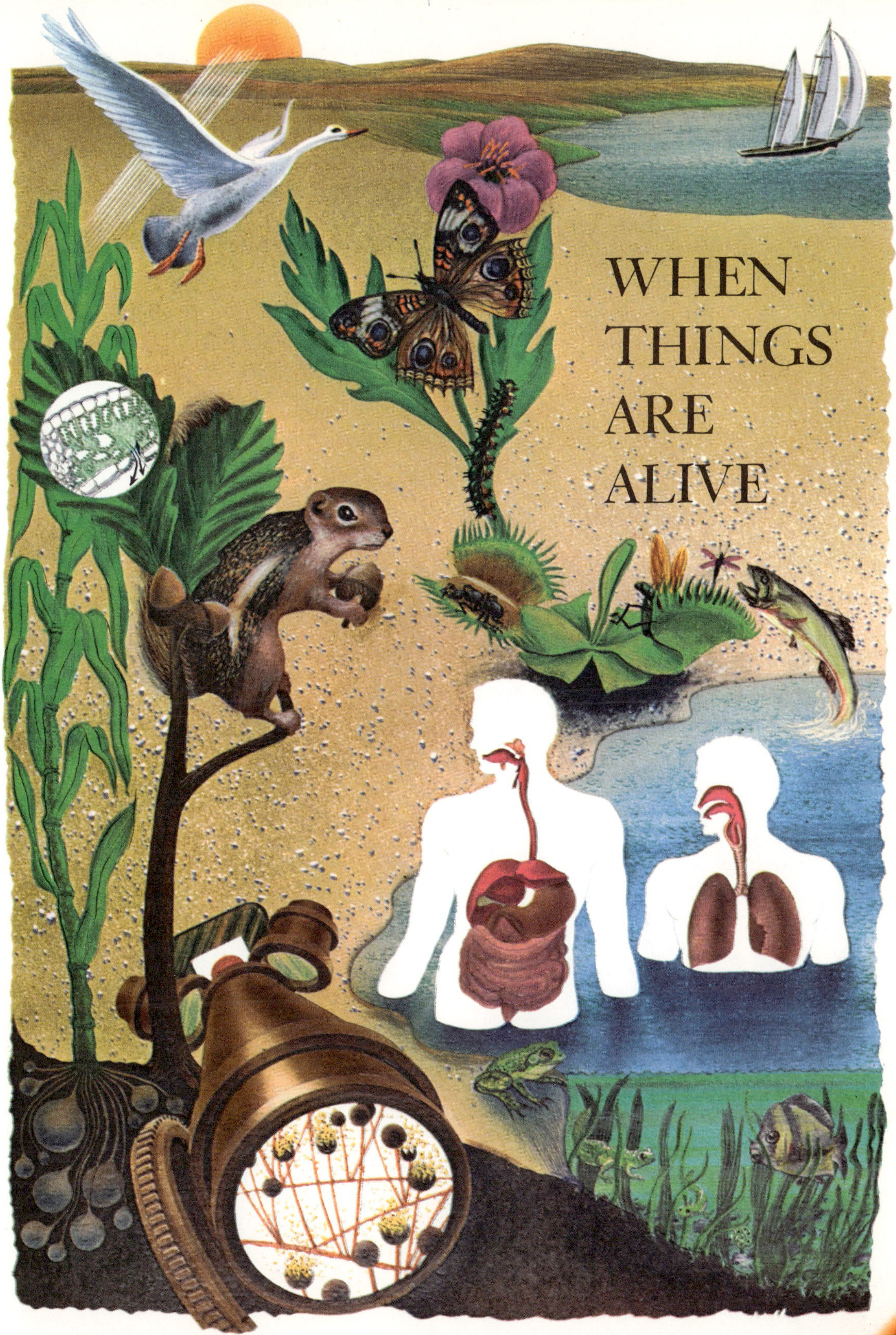

UNIT II

When Things Are Alive

CHAPTER 6. LIVING THINGS SHOW ACTION

CHAPTER 7. LIVING THINGS RESPOND TO OUTSIDE AND INSIDE CHANGES

CHAPTER 8. WE NEED FOOD TO LIVE

6

TARGET
How can we tell when things are alive?

Living Things Show Action

Fig. 41–1
This plane flies. Is it alive?

1. THE LIVING AND THE NONLIVING

"I slept like a rock last night," said Stephen. "I slept like a dead man," said his friend Jeff. These are colorful ways of speaking, and we can be sure that both boys had a good sleep. But you know, of course, that Stephen is not a rock, because he is ALIVE and rocks are not alive. We know that when Jeff slept, he stopped doing some things he does during the day. But he certainly continued to do many things, because he, too, is very much alive.

2. THE SIGNS OF LIFE

Have you ever watched someone who was asleep? Then you must have

Fig. 42–1

Each of the pictures shows motion. Can you tell the difference in the kind of motion? Which of these things is alive?

seen the person's chest move up and down. He probably changed his position many times. Perhaps he clenched his hand to make a fist. If a loose thread on the blanket tickled his ear, he brushed it away. When he felt cold, he pulled the blanket tightly around himself. When the alarm clock rang, perhaps he opened his eyes and then closed them again. When you tried to pull the blanket from him, he pulled it back. Finally, he got out of bed and the real world came back to him. He was fully awake.

Of course, rocks cannot do these things. And animals cannot do these things after they are dead.

3. NONLIVING THINGS SHOW MOVEMENT

Can a rock roll? Surely, but something must push it. Can a sailboat move? Yes, if the wind blows it. Can an automobile move? Of course, if its motor turns the wheels. Can a rocket enter space? Yes, if its jets blast it off the launching pad.

4. LIVING THINGS SHOW MOVEMENT

A squirrel can jump; a fish can swim; a bird can fly; humans can walk. These actions are controlled by the animals' bodies. A live geranium plant turns its leaves toward the light coming through the window. A live tree sends its roots toward a stream of water. A live ivy plant can spread its stems along a brick wall to a height of six or seven stories. These are examples of movement in plants and animals. How is the movement of a plant or animal different from the movement of a nonliving thing?

Fig. 43–1

How can you prove that the leaves turn toward light?

5. WHY DO LIVING THINGS MOVE?

Living things move for many reasons. A squirrel hops to get food. A cat runs to escape from a dog. The leaves of a tree turn to get light. Your friend in paragraph 2 pulls the blankets around himself to keep warm. Living things move to keep themselves alive and happy. They have to move in order to carry on their life functions: eating, fighting, getting light, getting air, getting water, etc. How is this different from the way a machine moves? A machine does not move so that it can carry on any life functions. It moves because someone built it that way. Only living things can move in order to carry on life functions.

6. EAGLES AND SNAILS

He's "as slow as a snail"! She's "as fast as an eagle"! Can't you just picture the people we are talking about?

Animals have many ways to move and they move with different speeds. The snail moves so slowly we can barely see it move. Slowly, he pushes along on a flap of muscle called a "foot." But he does get there! A deer can run at a speed of 45 miles an hour; the eagle flies at a speed of 60 miles per hour. The fastest bird, a type of swift, has been clocked at 106 miles per hour. The giant blue whale can swim at a speed of almost 25 miles per hour! How fast can you run? What is the world's record in the 100-yard dash?

7. THE MOVING PARTS

Animals move on land, in water and through the air. Simple animals have simple moving parts. The more complicated animals have either arms, legs, wings, fins or tails. The list below gives a few examples of organs of movement.

ANIMAL	ORGANS OF MOVEMENT
Microscopic animals	Hairs or "tails"
Starfish	Tube feet
Insects	Six legs; wings
Lobsters, spiders	Eight legs
Fish	Fins, tail
Snake	Body muscles and tail
Birds	Two rear legs; two front wings
Mammals	Four legs

YOU NOW KNOW

- Living plants and animals often show some kind of action which we call movement or LOCOMOTION (lo-co-MO-shun).
- Living things move in order to carry on their life functions.

UNDERSTANDING WHAT YOU HAVE READ

I. What are the two main ideas of Chapter 6?

1. Animals show more movement than plants.
2. Locomotion is one way to tell if a thing is alive.
3. After an animal is dead, it cannot move.
4. Living things move so they can carry on their life functions.

II. In which paragraph do you find the answer?

1. Does the flying of a rocket make it alive?
2. Which things needed by an animal make it move?
3. How do a tree's roots show that the tree is alive?
4. How do we know a sleeping puppy is alive?

III. Knowing What and Why

1. The rolling of a bowling ball is not a sign of life because
   a. it rolls faster or slower.
   b. someone gave it a push.
   c. rolling is not locomotion.
   d. the ball will not stop.
2. Leaves of living plants often face the window of a room
   a. to get warmth.
   b. to get light.
   c. for no reason. It's an accident.
   d. to get oxygen.
3. Sleeping people move so they can
   a. get air.
   b. receive stimuli.
   c. remove pressure against muscles.
   d. show signs of life.
4. Which of the following is *least* likely to cause a snake to move?
   a. fog
   b. noise
   c. odor
   d. sudden flash of light

IV. For the Home Laboratory

Get a healthy plant which is growing in a pot. Place a chalk mark on one side of the pot. Place the plant on the window sill with the chalk mark facing the window. Leave it for a few days. What did the leaves do? Now turn the pot so that the chalk mark faces into the room. What happens after a few days? Write a report to give to the class.

V. For the Sports-Minded

1. Using a good almanac or an athletics handbook, look up the world speed record for track runners and long-distance swimmers. What are the physical characteristics of such athletes?
2. Look up the world records of race horses. Are there differences in sex, weight and age? What was the record of the race-horse *Secretariat?*
3. Repeat your search for dog racing.

7

TARGET
What do living things do when changes take place around them?

Living Things Respond to Outside and Inside Changes

1. ANIMALS RESPOND

When you whistle at your dog, he comes running to you. We say he RESPONDS to you. In other words, he answers your whistle. A response is like an answer. When a bright light is flashed in your eyes, your eyes respond by closing. When you smell food cooking in your kitchen, you may respond by asking your mother for a sample. Your mouth may respond by "watering." If your finger touches a hot pan, it responds by pulling away quickly.

2. PLANTS RESPOND

You have just read about the responses of animals and you have seen them many times. Since you know that plants are alive, you have probably guessed that they also respond. You are right! When there is a change around the plant, the plant responds. For example, if you water one side of a bush, the roots will grow toward the watered side. Leaves of a plant will actually grow around a wall to get light.

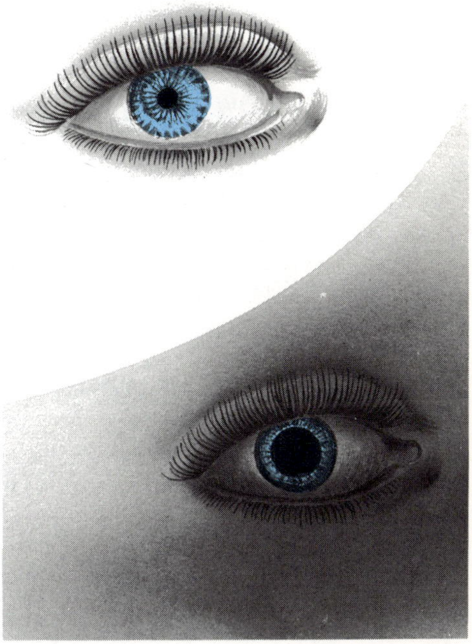

Fig. 45–1

Can you explain what is happening in this picture?

DO AND DISCOVER

Investigation 7: How do plants respond to light?

Fig. 46–1

Can you predict what will happen? Why? Can you suggest a way to study whether the color of the light will make a difference? (Hint: there are two ways.)

The surroundings of a plant or animal are called the ENVIRONMENT (en-VY-run-ment).

YOU NOW KNOW

- When there are changes in the environment, plants and animals show one or more responses to the change. They do something.

3. THE MEANING OF STIMULUS

Earlier, we saw that an alarm clock can wake a sleeping person. The new sound is a change in the environment. Any change in the environment is called a STIMULUS (STIM-you-lus). The stimulus is received by the plant or animal. The plant or animal then answers the stimulus with a response. Our sense organs are used to receive stimuli.

4. A STIMULUS BRINGS ON A RESPONSE

When the heat in your environment changes, we say that the temperature goes up or down. Your skin receives the stimulus of heat and you may respond by taking off your sweater. If it gets very hot, your skin may respond by sweating. If it gets very cold, your body may shiver and your skin may show "goose pimples."

YOU NOW KNOW

- A stimulus makes a living thing produce a response.
- The same stimulus may produce different kinds of responses.
- The responses are useful to the plant or animal.

5. SOME EXAMPLES OF RESPONSES

Stimulus	Response
Bright light	eyes blink, animal may run away
Dim light	eye openings get wider
Loud sound	heavy breathing, animal jumps, runs
Soft sound	animal soothed, may doze
Rough, sharp or hot	animal hurt—moans, screams
Object on skin	animal runs or fights
Good taste	animal licks lips, swallows, mouth waters
Bad taste	animal spits, runs away

6. ONE RESPONSE DEPENDS ON ANOTHER

You have probably been warned not to do exercise too soon after eating. This is an example of how one system in your body causes changes in another system. In this case, the action of your muscles can affect your digestive system to give you cramps. If you become very frightened, you may begin to run. Your muscles and skeleton begin to work. If you get angry, your heart begins to beat faster. These are examples of one system in your body responding to another system.

7. EMOTIONS CAUSE CHANGES IN YOUR BODY

"He turned red with anger."
"He turned white with fear."
"She blushed for shame."
"She laughed with joy."

The last word in each sentence is a feeling. Another word for feeling is EMOTION (ee-MO-shun). In each case the emotion does something to your body. Emotions affect your pulse, your heart, your muscles and your whole body. Emotions are sometimes useful. For example, the feeling of pain makes you pull away from a hot pan. But emotions, when not controlled, can hurt your heart and your nerves. Emotions can disturb sleep and digestion. People should learn early in life to control such emotions as anger.

YOU NOW KNOW

- Fear, excitement and anger are emotions.
- Emotions cause changes in many parts of the body.

UNDERSTANDING WHAT YOU HAVE READ

I. What are the two main ideas of Chapter 7?

1. Emotions are harmful.
2. Living things respond to changes around them.
3. Light is a very strong stimulus.
4. A stimulus brings a response.

II. Explanation please . . . answer in one or two sentences.

1. What do scientists call a change in the environment?
2. How does fear affect your body?
3. Why does a tree grow toward water?
4. What causes "goose pimples"?
5. What parts of our bodies receive stimuli?

III. Knowing What and Why

1. Dust in a person's nose makes him sneeze because
   a. the dust is a stimulus.
   b. the dust is a response to breathing.
   c. sneezing is natural.
   d. sneezing is a form of rapid breathing.
2. Some flowers close at night because
   a. it is cold.
   b. it is moist.
   c. it is dark.
   d. there is a lack of food.
3. When we get very cold, we sometimes shiver because
   a. shivering makes us exercise to get warmer.
   b. shivering is being afraid.
   c. shivering is a stimulus to protect ourselves.
   d. shivering is like "goose pimples."
4. An egg does not seem alive, but it can hatch into a chicken. Hatching is
   a. natural growth.
   b. a response to the stimulus of warmth.
   c. a stimulus for the young chick to break the shell.
   d. the chick's way of looking for a new environment.

IV. Using Your Judgment

1. Which is a stimulus? Which is a response?
   a. the smell of laundry soap
   b. a pistol shot
   c. the downward growth of roots
   d. bright headlights
   e. trembling
   f. a skunk's odor spray
   g. a sunflower's movement toward the sun
   h. a cat's attempt to scratch a dog
   i. the stiffening of a rabbit's ears
   j. the tickling of a baby's foot
2. Where you said "response" in Question 1, can you tell what stimulus might have caused it?

V. Why are these false?

Each of the following sentences is *false* because of the underlined word. What word could you put in its place to make each of the statements true?

1. Responses are <u>harmful</u>.
2. Shivering is caused by too much <u>heat</u>.
3. Sneezing is a <u>stimulus</u>.

VI. Go and Discover

Visit a local greenhouse. Ask the "keeper" to tell you how his plants respond to light, temperature and water. Write a report for your class.

DO AND DISCOVER

Investigation 8: How do plant seeds respond to water?

Fig. 49–1

Get a dozen fresh lima beans, two shallow soup plates and some absorbent cotton. Place six seeds in each dish between layers of cotton. Water the cotton in dish A every day. Leave dish B dry. Write a report of what happens and bring the report to class.

EXPERIMENT ANALYSIS

After you have presented your report, several classmates give you the challenges below. For each challenge, tell briefly how you might set up additional experiments to test their statements.

1. Jose: "Maybe the seeds in Dish B are too old to germinate."
2. Pamela: "It was too cold for the seeds to germinate."

8

TARGET
How do plants and animals take in food?

We Need Food to Live

Famine, Disease and Drought Ravage a State in India

40 Million Affected— Bihar Young Facing Permanent Injury

By J. ANTHONY LUKAS
Special to The New York Times

GAYA, India, April 14—The "loo," the hot, dusty wind of summer, has begun to blow across Bihar's parched acres.

As it sings eerily through the barren fields, villagers wrap shawls around their faces or huddle in the doorways of their huts. They all know the old Bihari saying: "The loo brings death."

This year the loo is truly a wind of death in Bihar. Its first gusts this month signaled the start of the testing time for the state, which has been wracked by famine and pestilence.

Fig. 50–1

1. FEAST, FAMINE AND LIFE

The newspaper clipping (left) tells the sad story of a FAMINE (FAM-in), a period of great shortage of food for large numbers of people. The result is that many people actually starve to death. Even those who live become ill and weak. Fortunately, such things do not occur often in modern times.

Perhaps you have heard a fat person say, "I'm starving myself these days. I'm bound to lose some weight." Or perhaps you have heard some mother say, "My girl is so skinny! She eats like a bird." Farmers often judge their soil as "rich" or "poor." What they really mean is how big, how fat or how plentiful the crop is that the soil produces.

These examples show that food has something to do with the health and the life of plants and animals. When there is plenty of the right kind of food, animals grow and stay healthy. When soil contains enough of the right kind of minerals, the plants grow tall and strong. In extreme cases, where there is no food at all, the plants and animals will die of starvation.

2. GETTING THE FOOD WE NEED

"Eating" is a common word and we all know what it means. Somehow, each animal has a way of getting food into its body. When we think

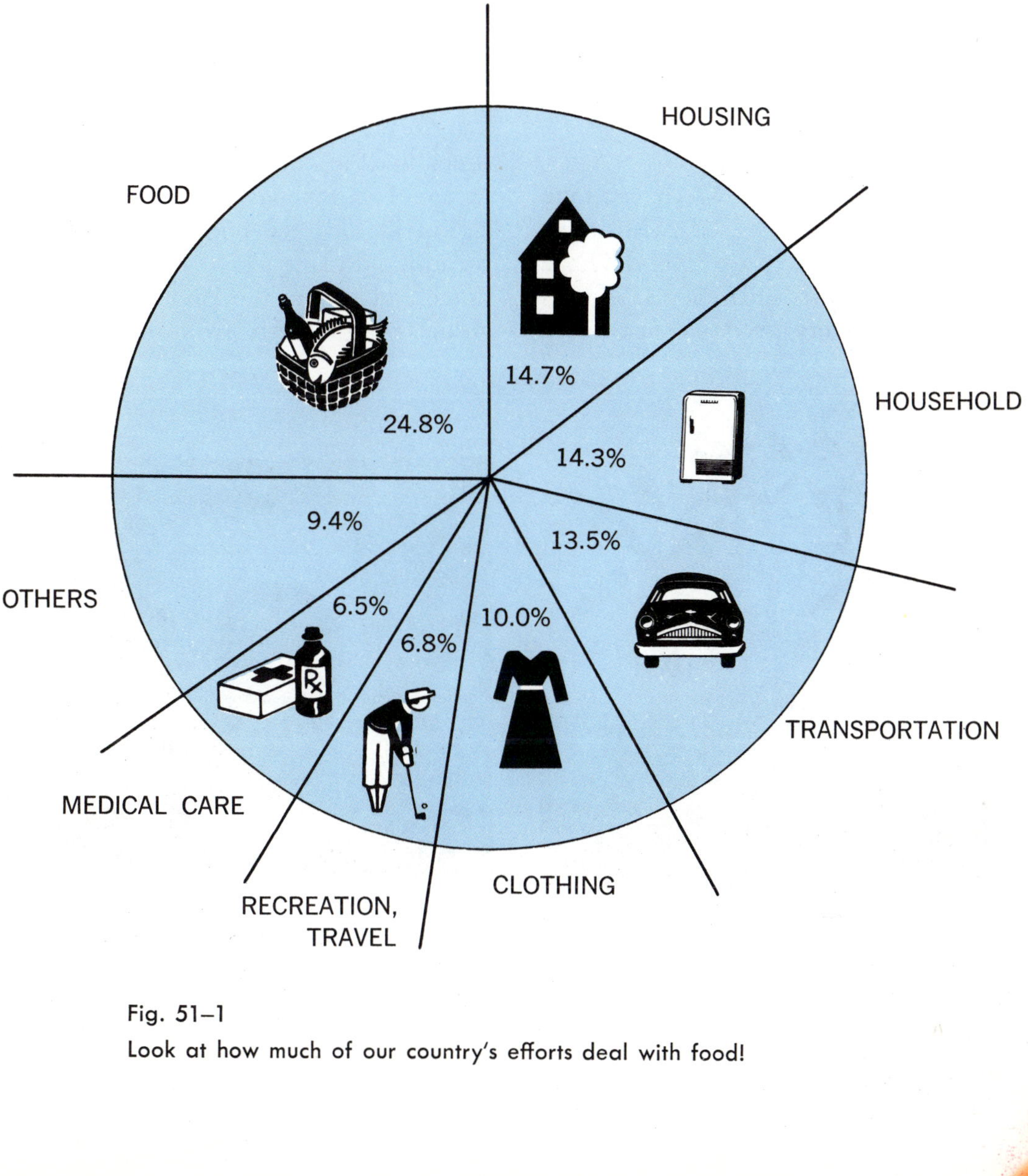

Fig. 51–1
Look at how much of our country's efforts deal with food!

of eating, we usually think of a mouth. We know that most animals have some kind of mouth through which the food enters the body. In the more highly developed animals, the hands, paws or claws grasp the food and place it into the mouth. Many animals have sharp teeth and powerful jaws. These help them cut, rip and grind the food into smaller bits before it is swallowed. The taking in of food is called INGESTION (in-JES-chun).

3. HOW PLANTS GET FOOD

Plants ABSORB (ab-SORB) or take in water from the soil with their roots. Soils which are "good" have chemicals called MINERALS, which help the plants grow. When plant roots absorb water, they also absorb any minerals which may be dissolved in the water. Sometimes we add minerals to the soil by using mixtures of chemicals called FERTILIZERS (FER-til-ly-zerz). Later you will learn that green plants can also manufacture their own food.

4. PLANTS AND ANIMALS WITHOUT MOUTHS

There are, of course, some very simple animals which do not have mouths. An example of such an animal is the tapeworm which sometimes lives in the body of a human being. The tapeworm absorbs food right through its body wall. The bread mold plant which you grew at home and saw in the laboratory can absorb food directly from the bread.

Fig. 52–1

Plant roots absorb water and dissolved minerals through thin root hairs.

Fig. 52–2

Plants can be fed artificially by adding chemicals to the soil. What are such mixtures called?

Tapeworm

Bread Mold

Fig. 53–1

Some animals and plants absorb food. They do not ingest food.

There are also the insect-catching plants which live on live insects. The Venus's-flytrap shown below is one such plant (Section C, below).

YOU NOW KNOW

- Plants and animals need food.
- The need for food is a sign of life.

Fig. 53–2

UNDERSTANDING WHAT YOU HAVE READ

I. What are the two main ideas of Chapter 8?

1. Food is needed for life.
2. Famines are dangerous.
3. Fertilizers are added to the soil to help plants grow.
4. All plants and animals have some way of getting food.

II. In which paragraph do you find the answer?

1. What materials in soil help plants grow best?
2. How does a tapeworm get food?
3. What do we call plant "food" that farmers add to soil?
4. Which plant eats insects?

III. Knowing What and Why

1. A person diets in order to
   a. lose weight only.
   b. gain weight only.
   c. gain or lose weight.
   d. avoid starvation.
2. Plants get their food
   a. by absorbing it from soil.
   b. from fertilizers.
   c. from other plants.
   d. by absorbing minerals and water and then making food.
3. Which of these absorb food directly?
   a. Venus's-flytrap
   b. bread mold
   c. ants
   d. bees
4. About how much of United States industry deals with food?
   a. one-fifth (1/5)
   b. one-quarter (1/4)
   c. one-half (1/2)
   d. two-thirds (2/3)

IV. How are the two words different?

1. ingestion—chewing
2. famine—starvation
3. fertilizer—soil water

V. For the Amateur Scientist

1. *Problem:* Does fertilizer help plants grow?

Get a small amount of fertilizer and two healthy geranium plants. Add about half a teaspoon of fertilizer to one of the plants once a week. Continue to water both plants with the same amount of water each day. Write a report of your results and bring it to class.

2. *Problem:* How do plant stems carry water to the leaves?

Get a fresh stalk of celery. Cut the end off about one inch from the bottom. Place it in a jar of water which you have colored with ink. Observe it for a few days. Write a report to explain what you see and bring it to class.

VI. Read all about it

In the library, read about *organic gardening*. Write a report, covering the following guidelines:

1. What are the claimed advantages?
2. Is it always better than "chemical gardening"?
3. Are all claims of organic gardening backed up by scientific tests?
4. Is organic gardening a fad?

9

TARGET
What is meant by digestion and circulation?

Food Must Be Changed and Delivered to All Parts of the Body

1. MAKING FOOD READY FOR USE

Have you ever heard of cooking food *twice?* This may sound silly, but it is what happens every time you eat a hamburger. The meat is cooked once before you eat it. Then, after you eat the hamburger, it is "cooked" a second time by your body. Your body "cooks" or makes ready for use all the food that you eat. This process of the body which makes food ready for use is called DIGESTION (di-JES-chun).

2. WHAT IS DIGESTION AND WHY IS IT NECESSARY?

Earlier we said that some animals chop, grind and cut food with their teeth. You will recall that these are methods which produce only physical changes. The molecules in the food remain the same. Foods contain many kinds of chemicals. Living things are able to use certain chemicals, but not others. To be able to get all the chemicals they need from foods, living things must have ways of changing the food *chemically* to simpler forms. These chemical changes are a part of digestion.

In the tapeworm, which has no mouth, and in plants which do not really "eat," digestion must still take place. Food must still be changed chemically before it can be used.

YOU NOW KNOW

- Digestion is a whole series of chemical changes.
- Plants and animals digest foods into simpler chemicals so that they can use them for life processes.

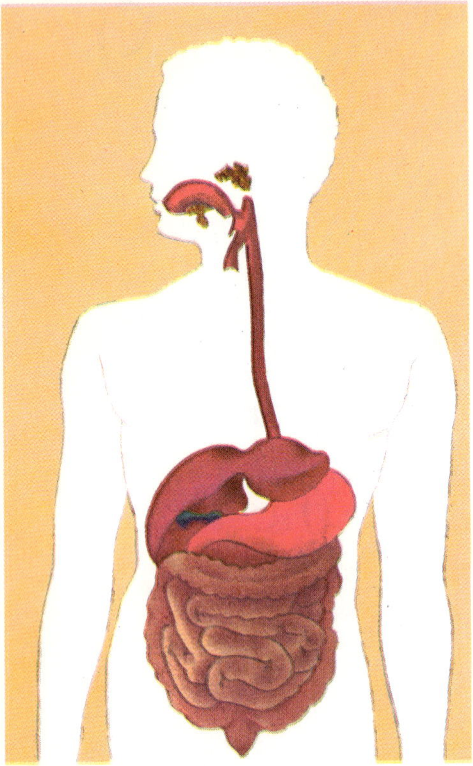

Fig. 56–1

Food is digested as it moves through most of the food tube. What is the advantage of a long digestive tube?

3. WHERE DOES DIGESTION TAKE PLACE?

In most animals, including human beings, the digestive changes take place in a digestive canal or food canal. Examine the diagram of a human food canal.

4. DIGESTIVE JUICES CAUSE CHEMICAL CHANGES IN FOOD

As the food moves through the food tube it is mixed and "bathed" in digestive juices. These chemicals are made in digestive GLANDS. The glands you know best are the SALIVARY (SAL-i-ver-ee) GLANDS in the mouth which make a juice called SALIVA (sa-LY-va). You have surely felt this gland working when your mouth "waters." Actually, the saliva made by the glands is not altogether water. Dissolved in the water are special chemicals which can cause chemical changes in certain foods that you eat.

YOU NOW KNOW

- Glands make chemical juices.
- These juices change food into simpler forms of chemicals in a food tube (canal).
- Digestion is a sign of life.

5. FROM HEAD TO FOOT

How long is a whale? There are some whales which are a hundred feet long. Some whales are three or four times as long as your classroom. The whale's mouth is in his head, of course. But the food he eats must be able to reach every part of his body —even the muscles in his powerful tail fin, about ninety-five feet away! How far is your big toe from your mouth? Perhaps five feet. Yet your big toes and your ears and your brain must be fed. All parts of a plant or animal must be fed.

6. DELIVERING THE FINISHED PRODUCT

You can see that it is not enough to digest food. It must also be moved or transported to all parts of the body.

Fig. 57–1

The stream carries logs. The blood carries digested food.

In many animals there is a system of branched tubes filled with a liquid called blood. This blood moves around and around in a circular path through the whole body. We call this flow of blood CIRCULATION (ser-kew-LAY-shun). Blood carries digested food to all parts of the body.

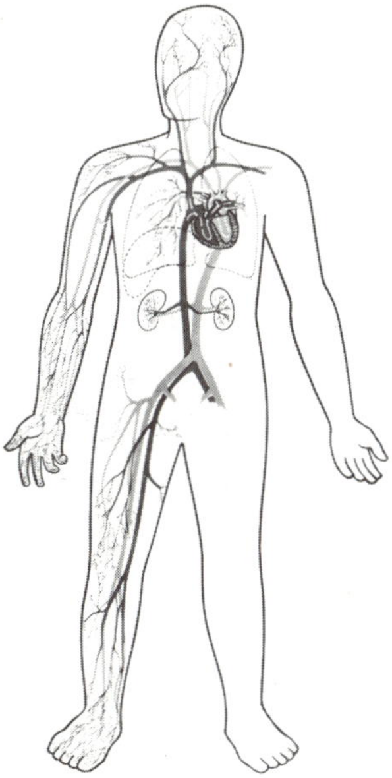

Fig. 57–2

The blood tubes carry digested food to all parts of the body. Why is this necessary?

7. FLOATING IN THE BLOOD TUBES

The blood in the blood tubes can be compared to the water in a river. Instead of logs, we can think of the small, simple molecules of digested food floating in the blood. This circulation carries the digested food molecules to parts of the body where they are needed—head, toes, brain, fingers—all parts.

YOU NOW KNOW

- Many animals have blood tubes in which blood circulates through their bodies.
- The blood carries dissolved and digested food molecules.
- Circulation is a sign of life.

UNDERSTANDING WHAT YOU HAVE READ

I. What are the two main ideas of Chapter 9?

1. Digestion changes food chemically.
2. Plants use simple foods.
3. The circulation of blood carries food to all parts of the body.
4. Tapeworms do not have a digestive system.

II. In which paragraph do you find the answer?

1. What juice is in your mouth when it "waters"?
2. What tube carries food from the mouth to the stomach?
3. How does blood flow through the body?
4. What is digestion?
5. How can tapeworms live without "eating"?

III. Knowing What and Why

1. Digestion acts on food to
   a. change it into simple molecules.
   b. complete ingestion.
   c. change it to blood.
   d. move it to the stomach.
2. The glands in your mouth
   a. make saliva.
   b. cause physical changes in food.
   c. give you an appetite.
   d. make taste possible.
3. Eating and chewing do not cause digestion because
   a. they take too short a time.
   b. the food does not go far.
   c. these are physical changes.
   d. these are chemical changes.
4. Circulation is necessary
   a. because most food molecules cannot be used in the form that we eat them.
   b. because digested foods are more nourishing.
   c. because food must move rapidly through the food canal.
   d. to move simple food molecules.

IV. On the Ladder of Understanding

1. Molecules contain ______.
2. Elements combine to form ______.
3. Sneezing is a kind of ______.
4. One stimulus that causes shivering is ______.
5. The simplest form of a chemical is called ______.

V. Unscramble the letters.

The answers are at the bottom of the page.

1. FZETRILIRE
2. OTGIDIESN
3. ONIGSTIEN
4. DLNAG
5. ALSIVA

(Answers are upside down.)

1. fertilizer 2. digestion 3. ingestion
4. gland 5. saliva

10

TARGET
Why is breathing a sign of life? What wastes are formed by living things?

Using the Food and Getting Rid of the Wastes

1. "IS HE STILL BREATHING?"

This is the most common question asked when we want to know whether someone is alive. It is easy to see your parakeet, your puppy or your friend breathing. You can see his chest grow larger and smaller, as he takes in air and gives out air. We call this BREATHING. In the pictures below, you can see how different living things breathe.

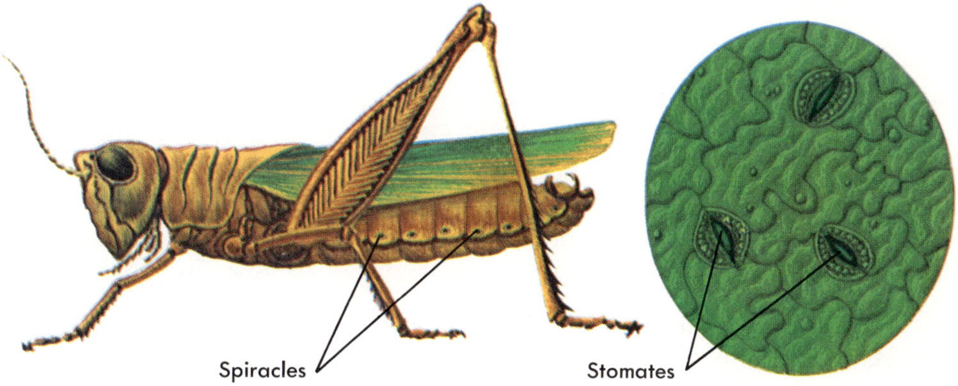

Fig. 59–1
Insects breathe through openings on the side of their bodies called *spiracles*. Plants breathe through stomates on the underside of their leaves.

2. HOW ARE WASTES FORMED?

In the coal furnace, part of the fuel cannot burn. It is left over as a waste, called ashes. When animals eat food, some solid parts are not digested. This is left as solid waste. Water is also left over as a waste.

3. WASTES RELEASED BY BREATHING

Chemists know that carbon dioxide forms if you burn materials containing carbon. This is called OXIDATION (ox i DAY shun). Food contains carbon in its chemical compounds. You breathe out carbon dioxide. Look at the diagram below.

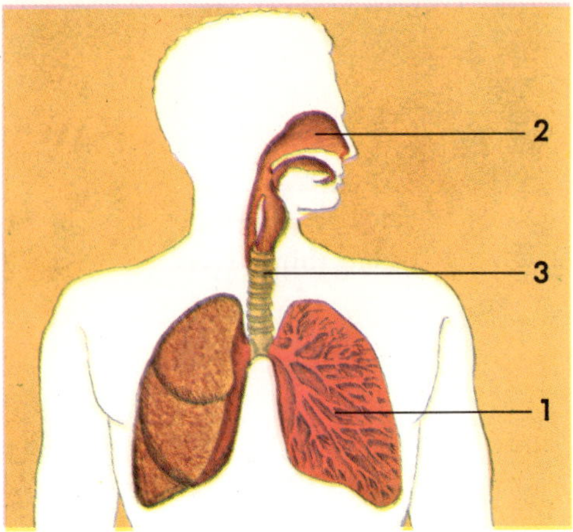

Fig. 60–1

We breathe to obtain air and its oxygen. Can you name the numbered parts?

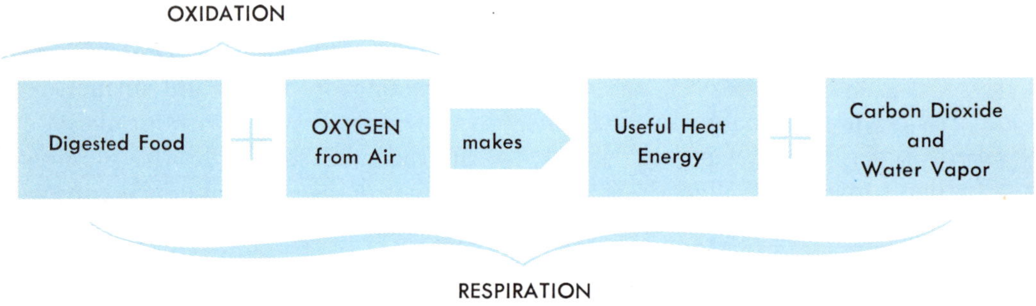

The oxidation of food is called RESPIRATION (res-puh-RAY-shun). The two wastes formed in respiration are two gases you have studied before: *carbon dioxide* and *water vapor.*

4. "BOY, WAS I SWEATING!"

How often you have heard this remark after coming out of the gym! Of course, the harder you play, the more energy you need, and the more food you "burn" through oxidation. You breathe faster and your skin gives off SWEAT (SWET). Another name for sweat is PERSPIRATION (per-spuh-RAY-shun). Perspiration is a mixture of water, salt and other wastes. Sweating is one way your body gives off waste.

5. GETTING RID OF WASTES

In living things all life actions produce wastes. These wastes must be removed from the body of the plant or animal to keep it working smoothly and in good health. The removal of wastes from living things is called EXCRETION (ex-KREE-shun).

6. WATER MEANS LIFE

One of the greatest fears of lost people is that they will die of thirst. The shipwrecked sailor on a lonely island; the flyer forced down in the desert; the people on rafts and lifeboats all search for the same "treasure": water! Why? All land animals must have a balance of water taken in and water given off. The amount of water in the body is regulated in two ways:

1. by urination: urine is formed in the KIDNEYS (KID-neez) from water in the blood. The urine is stored in the BLADDER.

2. by sweating through the pores of the skin.

If too much water is lost, the body becomes DEHYDRATED (dee-HY-dray-tid). Cramps and weakness are early signs. Unless water is taken in, death results.

7. HOW MUCH SWEAT?

DO AND DISCOVER

Investigation 9: How much sweat is given off by the skin?

What You Do

1. Place your thumb inside a dry, 6-inch test tube. Seal the top around your finger with adhesive tape. Keep the tube on for thirty minutes while you sit *quietly*, reading or watching TV.

2. Remove the test tube and empty the sweat into a small graduated cylinder.

3. Repeat the experiment and exercise vigorously for about ten minutes while you collect the sweat.

What You See

1. How many mililiters (ml) are formed in Step 2? In Step 3?

2. What is the color of the sweat?

Analyzing the Results

1. Explain why different amounts of sweat formed.

2. Would smearing grease or cold cream on your thumb make a difference? Explain.

3. How could you collect sweat from your entire arm?

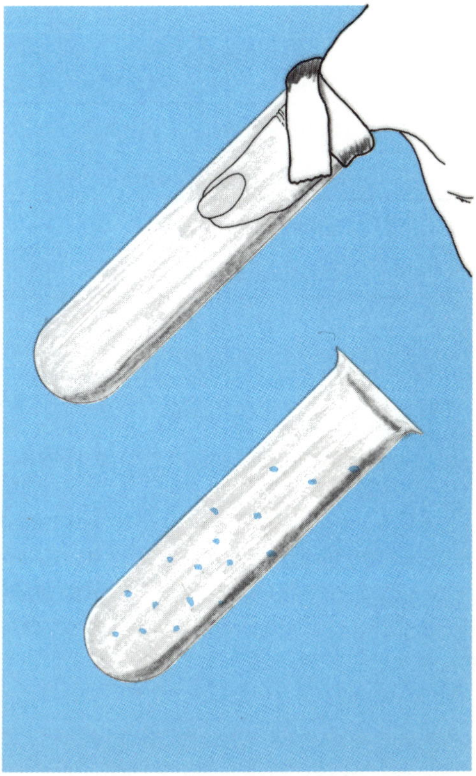

Fig. 62–1
Exercises use energy. What is the result? Why should one shower after exercising?

THE KINDS OF WASTES

Waste	What It Is
Solid waste	Undigested solids
Perspiration (sweat)	Water, salt and urea
Urine	Water, salt and urea
Exhaled air	Carbon dioxide and water vapor

YOU NOW KNOW

- Living things need oxygen for respiration.
- Living things form wastes after using food.
- Respiration is a sign of life.
- Excretion is a sign of life.

UNDERSTANDING WHAT YOU HAVE READ

I. What are the two main ideas of Chapter 10?

1. Most living things take in air to get oxygen.
2. Perspiration is necessary for health.
3. Respiration and the use of food make wastes.
4. Excretion follows breathing.

II. In which paragraph do you find the answer?

1. What is perspiration?
2. What chemical action produces energy in living things?
3. What are the breathing organs of insects called?
4. Is all food that we eat digested?

III. Knowing What and Why

1. Breathing means
   a. taking in air.
   b. oxidizing food.
   c. creating urea.
   d. creating carbon dioxide.
2. Urine contains water and
   a. oxygen.
   b. carbon dioxide.
   c. urea.
   d. undigested solids.
3. Oxygen is removed from the air in the
   a. windpipe.
   b. throat.
   c. stomates.
   d. lungs.
4. Respiration is a chemical process which uses
   a. ingestion.
   b. excretion.
   c. oxidation.
   d. perspiration.

IV. Fill in the missing letters.

1. __ __ GESTION: taking in food
2. EX __ __ __ TION: giving off wastes
3. O __ I __ ATION: burning food for energy
4. ST __ __ ULU __: a change in the environment that causes a response

V. Put on your thinking cap.

1. Why are rubber gloves often uncomfortable?
2. Why should cold creams and other cosmetics be washed from the face before going to sleep?

VI. Complete the story.

Use the words at the right to complete the sentences.

To get energy from food, we must (1) it. We take in air by (2) and then remove the useful element, (3). The wastes formed in this process are (4) and (5). If I were a grasshopper I would breathe with (6). Since I am human, I breathe through my (7) and send air to my lungs through my (8).

breathing
nitrogen
oxygen
water vapor
oxidize
spiracles
windpipe
carbon dioxide
stomates
nostrils

11

TARGET
How is growth a sign of life?

Growing and Living

1. BIG AND LITTLE—OLD AND YOUNG

Have you ever seen a picture of yourself when you were a baby? How much did you weigh? Do you remember your dog when he was a pup? Or your cat when she was a kitten? Can you remember the big tree on your street when it was first planted? What has happened to it since then? Perhaps your mother has shaken her head and wondered how she can keep buying you trousers fast enough to keep up with your growth. If you are a girl, you will probably need to change your dress size next year. How does this all happen?

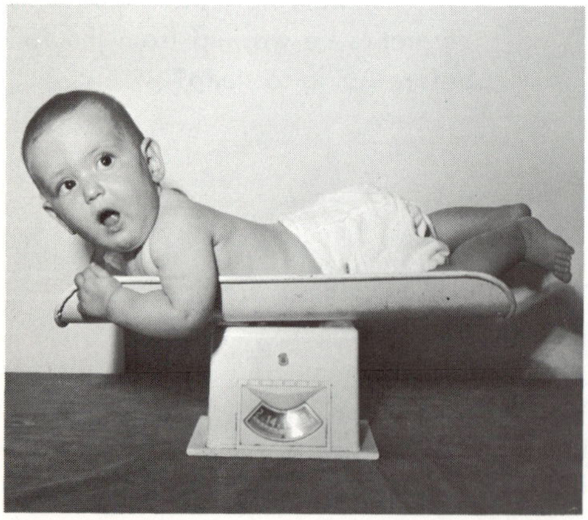

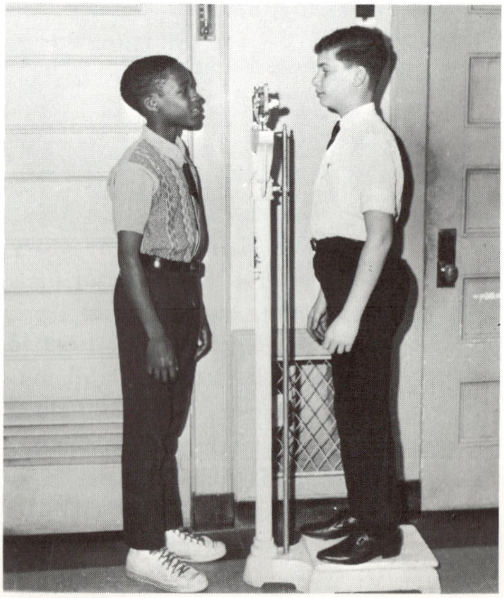

Fig. 64–1
How long will it take the baby to reach the weight of one of these boys?

2. GROWTH IS A SIGN OF LIFE

The examples of growing that we read about in paragraph 1 are only part of the whole story. We see many other examples of growth all around us. One haircut in your life is not enough—you need hundreds. Cutting your nails once is not enough. You must cut your nails quite often. Many men shave 365 times a year! Do you worry about the peeling skin on your sunburnt back? Of course not! You know that you will grow new skin very soon. Last week's cut finger is soon healed. After a short period, new flesh and new skin appear. Last year's broken wrist is only a memory because the broken ends of the bone were mended with new bone and your wrist is as good as before. The hedges you saw pruned (cut) last week have grown back again and need another cutting.

All of these are examples of the wonderful life process called GROWTH.

YOU NOW KNOW

- All living things show some kind of growth.
- Growth is a sign of life.

3. DIGESTED FOOD BECOMES LIVING MATTER

You have already learned that living things can form new skin, new bones, new leaves, new stems. It follows that this new material must be made from something. Yes, you have guessed right! New living matter comes from digested food. Living things are able to cause chemical changes in the molecules of the digested food so that the food becomes parts of their own bodies.

4. CHEMICALS ARE CHANGED INTO LIVING MATTER

Chemists have studied the composition of living matter. One kind of compound found in living things but never in nonliving things is called PROTEINS (PRO-teenz). Digested food must be turned into proteins. Our bodies also contain other chemicals. For example, our blood is rich in the element iron. Our bones are rich in calcium.

The living process by which plants and animals change food into living matter is called ASSIMILATION (a-sim-i-LAY-shun). Assimilation turns digested food into living matter for growth and repair.

YOU NOW KNOW

- Digested food is turned into living matter by assimilation.
- Assimilation makes growth possible.
- Assimilation repairs broken and damaged parts.
- Assimilation is a sign of life.

5. PLANTS SHOW GROWTH

Look back at page 24. You saw that plant cuttings can grow new roots. Tomatoes, apples, and corn are small in the early part of the season. As they assimilate sugar, starch, and protein, they grow bigger. Trees grow from tiny saplings to huge trees forty, fifty, or one hundred feet tall. Once again, food is changed by assimilation into proteins and other chemicals. Plants store food in their roots, stems and leaves. Tree trunks store special materials to form wood and bark. Trunks grow tall and give out many branches. Their leaves grow large. Growth is a life function of all plants.

Fig. 66–1

This giant redwood tree was once a young sapling. Assimilation of food made it tall and thick. Explain how.

THE YOUNG AND THE OLD

pups	dogs, seals
kittens	cats, leopards
cubs	bears, lions
goslings	geese
ducklings	ducks
fawns	deer
fry	fish
calves	cattle, elephant, whale
lamb	sheep
colt	horse
seedling	plant
sapling	tree

What factors control the growth of the young? In each group above, how do the parents contribute to the growth of their young?

UNDERSTANDING WHAT YOU HAVE READ

I. What are the two main ideas of Chapter 11?

1. Animals gain weight as they grow older.
2. Growth is a sign of life.
3. Some parts of our bodies grow back when damaged or removed.
4. Food is changed to living matter by chemical action.

II. Explanation please . . . answer in one or two sentences.

1. How do cuts in skin heal?
2. What process changes food to flesh and bone?
3. What chemical is needed to form bones?
4. How does a tree trunk grow thick?

III. Knowing What and Why

1. Proteins are chemical compounds which
   a. cause oxidation.
   b. are found only in living things.
   c. are waste products after respiration.
   d. are similar to carbon dioxide.
2. Growth is a life function that needs
   a. sun.
   b. excretion.
   c. food.
   d. ingestion.
3. If two plants are growing in the same soil and get the same amount of light and water, we can speed up the growth of one of them by adding
   a. carbon dioxide.
   b. oxygen.
   c. urea.
   d. fertilizer.
4. Assimilation means changing food to
   a. living matter.
   b. wastes.
   c. simple molecules.
   d. heat energy.

IV. For the Amateur Scientist

1. Ask a teacher who has a first grade class to give you the weights of the children in her class (without names).

   Collect the weights of the boys and girls in your own class (without names).

   Now get the *average* weight of the first grade class and compare it with the *average* weight of a seventh grade class.

   HINT: To get the average, add all the weights. Then divide by the number of children.

V. On the Ladder of Understanding

Protein, which is a compound, is made of many chemical ___ (1) ___.

Assimilation is a ___ (2) ___ change in the composition of food.

Calcium and iron are chemical ___ (3) ___.

The chemical combination of oxygen with carbon is called ___ (4) ___ and makes a compound called ___ (5) ___.

VI. For the Collector

Write to: World Health Organization
United Nations
New York, N.Y. 10017

Ask for: Booklets about feeding underdeveloped nations.

12

TARGET
How do plants and animals "live forever"?

Living Things Produce New Generations

1. THE NEXT GENERATION

Not only do living things grow large; they also grow old. And of course you know that sooner or later living things die. And yet there are always new plants and animals. As a matter of fact, we know that there are more plants and animals today than at the time of Columbus or Washington. The reason is that living things produce young. We sometimes say that plants and animals "give birth." This process is called REPRODUCTION (ree-pro-DUK-shun). Plants and animals leave a new generation. In this way they "live forever."

Fig. 69–1
The young and the old.

2. POPULATION GROWS

Scientists recently guessed that in forty years the human population of the world will *double*. As you know, when a dog has a litter of puppies, it is usually not just two puppies. In other words, the father and mother are not just replaced. Generally, a litter consists of many puppies. This is true of human beings too. Although only one child is usually born at a time, most families have several children. Some families have TWINS (two children that are born at the same time) or TRIPLETS (three) or even QUADRUPLETS (four). There are even rare cases of QUINTUPLETS (five)!

3. REPRODUCTION IN STAGES

Have you heard of the TADPOLE, a tiny creature that lives in water in the spring? Did you know that a baby tadpole does not grow up to be a "papa tadpole," but a frog? This is a special kind of growth, in which the young look very different from the old.

Another example is the crawling

Fig. 69–2
Some living things look very different when they are young.

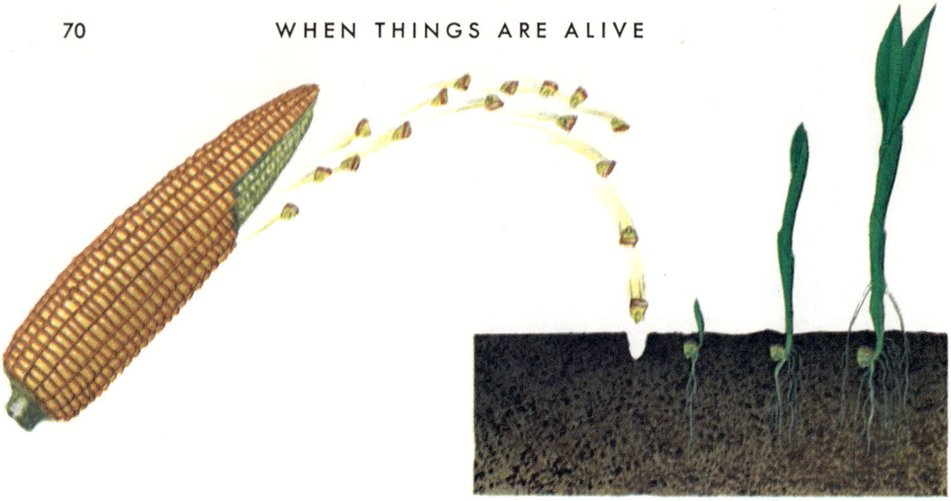

Fig. 70–1

Seeds grow into plants. Examine Figures 52–1 and 52–2. How are those pictures related to this diagram, above.

CATERPILLAR, with fuzzy body and many legs, which grows to be a beautiful butterfly with wings.

4. REPRODUCTION WITH SEEDS

Did you know that one pound of corn seeds, properly planted, can grow hundreds of pounds of corn-on-the-cob? Did you know that the Pilgrims brought seeds of wheat to Plymouth to plant new crops? Seeds are made by plants for reproduction. Apple seeds grow into apple trees; pumpkin seeds grow into pumpkins.

Fig. 70–2

Bread-mold branches make spores, which are scattered in the air.

5. REPRODUCTION WITH SPORES

In Unit I, Chapter 4, we saw how mold can grow on bread. The living mold "shoots out" tiny, invisible SPORES. When other spores land on bread, jelly, cheese, or fruit, they grow to be a new mold. By keeping food covered, we keep the spores away and stop food from spoiling.

6. REPRODUCTION WITH EGGS AND SPERMS

Many plants and animals have two sexes, male and female. Males produce SPERMS and females produce EGGS. When a sperm and an egg combine, the new bit of living matter

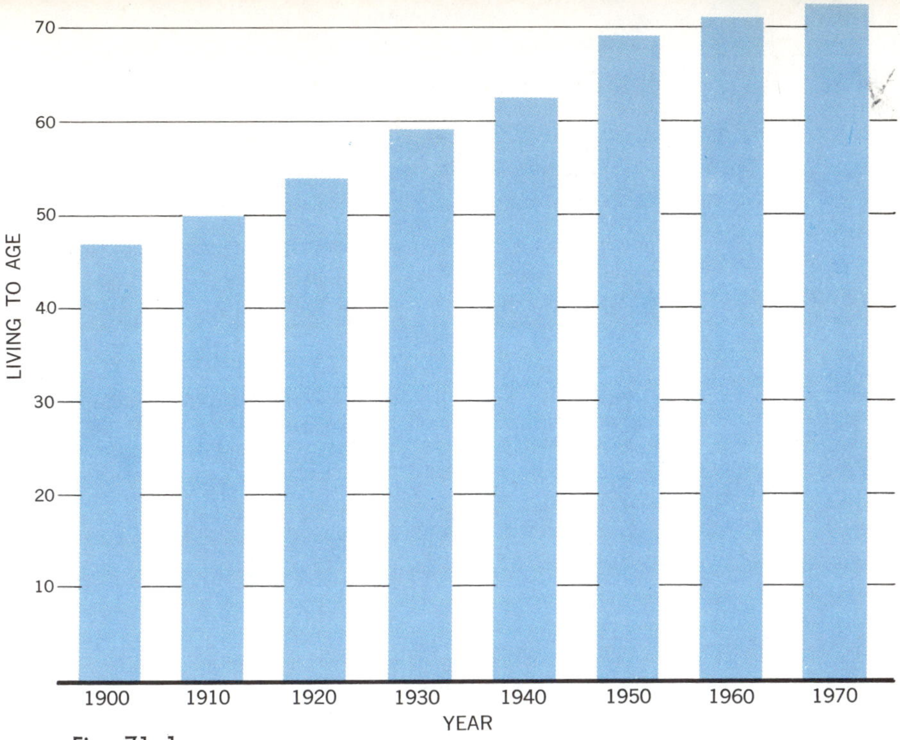

Fig. 71–1
Life expectancy in the U.S. What does the graph show? Give at least two reasons for the changes shown.

grows into a new living thing. This is called reproduction through SEX. Later, you will learn more about sexual reproduction.

Fig. 71–2
Mushrooms also reproduce by spores. The spores fall onto the soil and grow new mushrooms.

7. LIVING LONGER

Unless disease or famine or disasters such as forest fires or volcanoes destroy living things, populations keep increasing. With human beings and domestic stock (cows, sheep, chickens), we have found ways to stop disease and protect the young. Therefore, humans and domestic animals live longer lives than they used to. We say we have increased LONGEVITY (lon-JEV-i-tee). Study the chart above. Can you tell what has happened to the longevity of humans in the last fifty years?

YOU NOW KNOW

- Living things reproduce young like themselves.
- Reproduction is a sign of life.

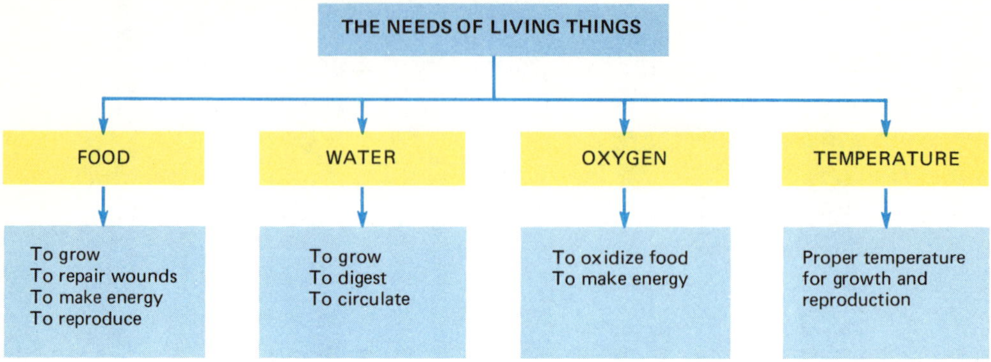
THE NEEDS OF LIVING THINGS
FOOD
WATER
OXYGEN
TEMPERATURE
To grow
To repair wounds
To make energy
To reproduce
To grow
To digest
To circulate
To oxidize food
To make energy
Proper temperature
for growth and
reproduction

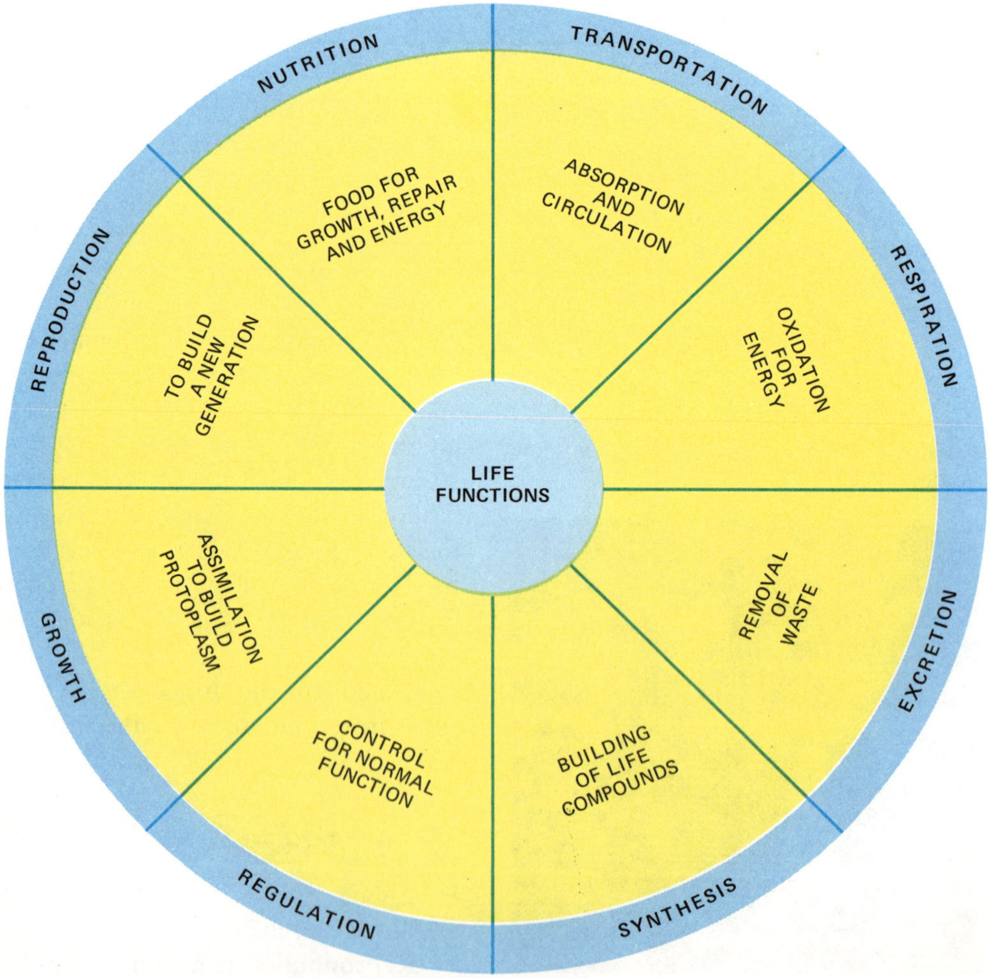
LIFE FUNCTIONS
NUTRITION
FOOD FOR GROWTH, REPAIR AND ENERGY
TRANSPORTATION
ABSORPTION AND CIRCULATION
RESPIRATION
OXIDATION FOR ENERGY
EXCRETION
REMOVAL OF WASTE
SYNTHESIS
BUILDING OF LIFE COMPOUNDS
REGULATION
CONTROL FOR NORMAL FUNCTION
GROWTH
ASSIMILATION TO BUILD PROTOPLASM
REPRODUCTION
TO BUILD A NEW GENERATION

UNDERSTANDING WHAT YOU HAVE READ

I. What are the two main ideas of Chapter 12?

1. Spores produce young.
2. Living things produce a new generation.
3. Population keeps increasing.
4. Reproduction is a sign of life.

II. Explanation please . . . answer in one or two sentences.

1. How does population increase?
2. Why is reproduction needed by living things?
3. How do mushrooms reproduce?
4. Why do humans and farm animals live longer now than years ago?

III. Knowing What and Why

1. Longevity means
   a. more young are produced.
   b. we live longer.
   c. freedom from disease.
   d. reproduction in stages.
2. Reproduction by sex requires
   a. seeds and spores.
   b. assimilation and longevity.
   c. growth and oxidation.
   d. sperms and eggs.
3. Tadpoles are to frogs as caterpillars are to
   a. forests.
   b. ponds.
   c. butterflies.
   d. molds.
4. What is the approximate age to which a man lives? (See Fig. 71–1)
   a. 72 years
   b. 68 years
   c. 54 years
   d. cannot tell from the chart

IV. Fill in the Blanks

A

1. Perspiration is an example of ________.
2. Spores are a part of the process known as ______.
3. Oxygen is necessary to carry on ______.
4. The body prepares food for use by the process known as ______.
5. Growth is made possible by ______.
6. Blood is carried to all parts of the body by ______.
7. Eating is ______.

B

a. digestion
b. respiration
c. assimilation
d. reproduction
e. circulation
f. excretion
g. ingestion
h. longevity
i. saliva

V. The first letter of each word spells something you eat.

1. Taking in air ______
2. Producing young ______
3. Giving off waste ______
4. Changing food to living matter ______
5. Changing food to simple molecules ______

What are the first letters of each word?

___ ___ ___ ___ ___

VI. Put on Your Thinking Cap

1. Life expectancy for American Indians is only 47 years. How can you explain this?
2. Do you think this might change by the year 2000? Explain.

Glossary: What Does It Mean?

absorb	ab SORB	To take in.
assimilation	a sim i LAY shun	Life process by which plants and animals change food into living matter.
bladder	BLAD-ur	A muscular sac that stores urine before it is excreted.
breathing	BREETH ing	Taking air in and letting it out.
circulation	ser kew LAY shun	The flow of blood and other materials in the body.
dehydrated	dee-HY-dray-tid	Removal or loss of water.
digestion	di JES chun	Process by which food is changed into simpler chemicals, which can then be used for life processes.
emotion	ee MO shun	A feeling.
environment	en VY ron ment	The surroundings of a plant or animal.
excretion	ex KREE shun	Removal of wastes from living things.
famine	FAM in	A period of great shortage of food for many people.
fertilizer	FER ti ly zer	A mixture of chemicals put into the soil to feed plants.
gills		Parts of fish used for breathing.
glands		Organs of the body producing chemical juices.
ingestion	in JES chun	Taking in of food.
locomotion	lo co MO shun	Movement.
longevity	lon JEV i tee	Length of life.
kidneys	KID-neez	A pair of organs that remove wastes from blood and make urine.
perspiration	per spuh RAY shun	A waste excreted through the skin.

proteins	PRO teens	Compounds found in living things—contain nitrogen.
reproduction	ree pro DUK shun	Process of giving birth to young of the same kind.
respiration	res puh RAY shun	Oxidation (burning) of food for energy.
response	re SPONS	An action following a stimulus.
saliva	sa LY va	Juice produced in the mouth by the salivary glands.
salivary glands	SAL i ver ee	Glands in the mouth which produce saliva.
stimulus	STIM you lus	A change in the environment.
stomates	STO mates	Tiny holes on the bottom of plant leaves used for breathing.
sweat	SWET	A liquid waste, another name for perspiration.
urine	YUR-in	Liquid waste from animals.

UNIT III

The body on the left shows the nervous system. The body on the right shows the blood system. The circles show certain cells as they look under a microscope. The circles in the pond show three types of single-celled life. The brick wall shows cells as the building blocks of life.

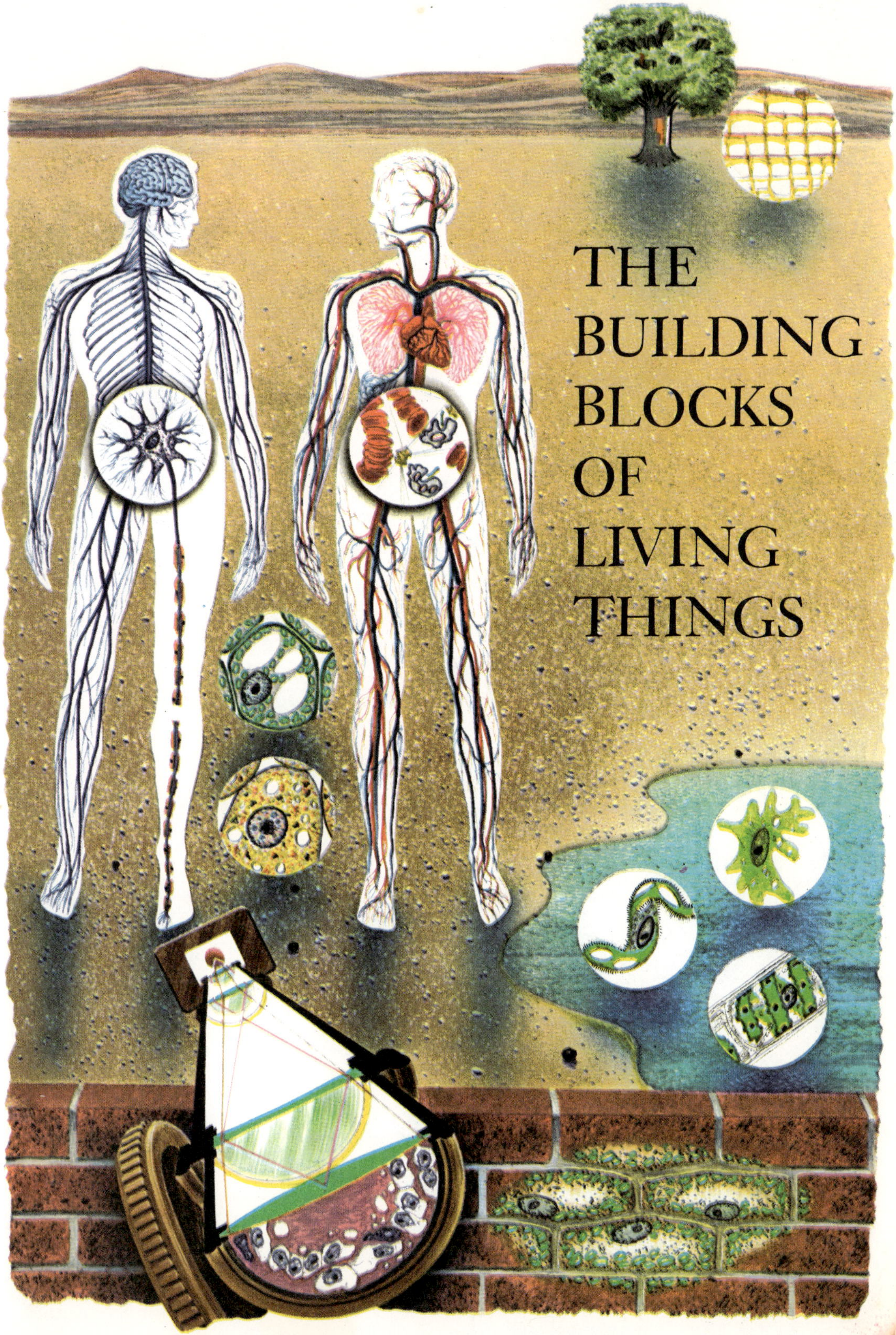
THE
BUILDING
BLOCKS
OF
LIVING
THINGS

UNIT III

The Building Blocks of Living Things

CHAPTER 13. THE MICROSCOPE—WONDERFUL TOOL OF SCIENCE

CHAPTER 14. THE MICROSCOPE AND THE UNSEEN WORLD

CHAPTER 15. A LOOK INSIDE THE CELLS

CHAPTER 16. GETTING THE WORK DONE INSIDE A CELL

CHAPTER 17. TEAMWORK AMONG CELLS

CHAPTER 18. TISSUES IN COMMAND AND CONTROL

CHAPTER 19. TEAMWORK AMONG TISSUES

13

TARGET
What is a microscope and how does it work?

The Microscope—Wonderful Tool of Science

1. WHEN THINGS ARE SMALLER THAN THE EYE CAN SEE

Have you ever seen a watch-repair man at work? He often wears a glass object over his eye. The glass object is a MAGNIFYING GLASS. When he looks inside a watch, he places the magnifying glass in front of his eye. A LENS, which is a specially prepared piece of glass, then magnifies, or enlarges, the tiny wheels in the watch. Such lenses were invented by the ancient Chinese people. The simplest lens, shown in Fig. 81–1, is a single glass lens. You can use it to examine the wings of an insect, the pores in your skin, the inner parts of a flower or the eyes of a lobster. Also, you can enlarge a whole animal, such as a tiny flea or a termite. A lens is a simple MICROSCOPE (MY-kruh-skope). *Micro* means small; *scope* means to see. A microscope is an instrument used to see small objects.

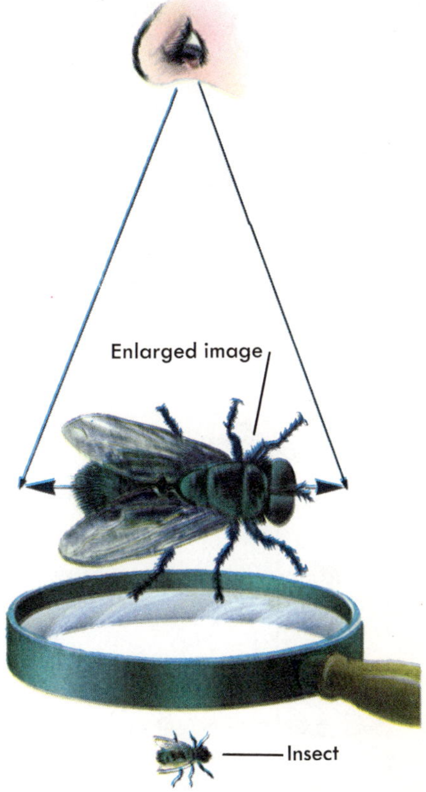

Fig. 81–1
A magnifying glass enlarges the image of a tiny insect.

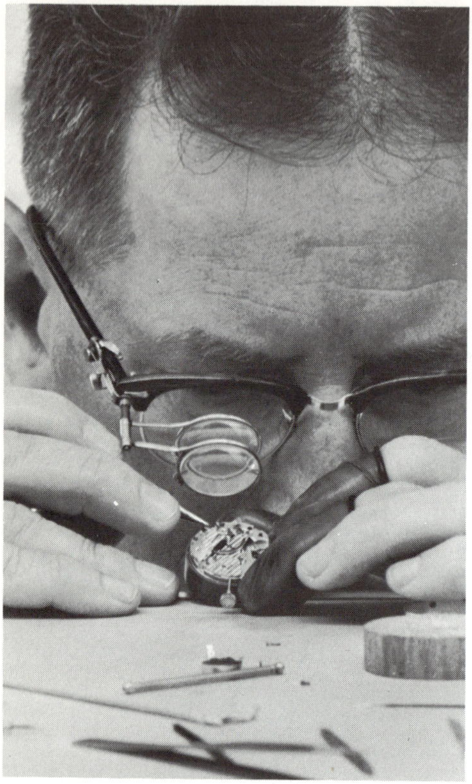

Fig. 82–1

A watchmaker uses a simple lens to see the tiny wheels in your watch.

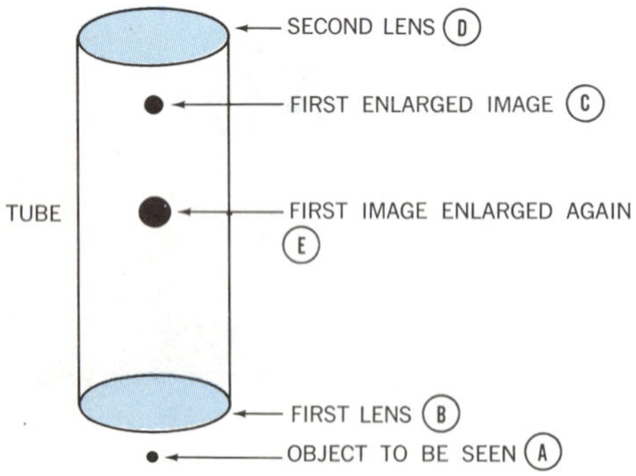

Fig. 82–2

Can you explain how this instrument works?

2. TWO LENSES ARE BETTER THAN ONE

Suppose you wanted to enlarge the image of an object still more than you could with one lens. Scientists long ago found a very simple way. They used two lenses, placing one behind the other inside of a tube. (See Fig. 82–2.)

The object to be examined (A) is magnified by the first lens (B). This gives us a larger image (C). The second lens (D) magnifies the image (C) to a still larger image (E)!

YOU NOW KNOW

- A lens is a specially prepared piece of glass that can magnify small objects.
- Two lenses placed one behind the other can give greater magnification than one lens alone.

3. THE IDEA OF THE COMPOUND MICROSCOPE

In your work in chemistry, you learned that a chemical compound is a substance that contains at least two elements. A microscope that combines two lenses, as explained in paragraph 2, is called a COMPOUND MICROSCOPE. It is easy to figure out *how much* a microscope can magnify by multiplying the power of one lens by the power of the other. For example, suppose Lens B in Fig. 82–2 magnifies *five* times (5X) and Lens D magnifies this image *ten* times (10X). We then multiply the power of Lens B by the power of Lens D. The Image E which we see is enlarged 5 times 10, or 50 times. Many school microscopes have lenses that enlarge 10X with the first lens and 10X with the second lens. Your eye will then see the object 10 times 10 or 100 times as large as its true size.

Fig. 83–1a

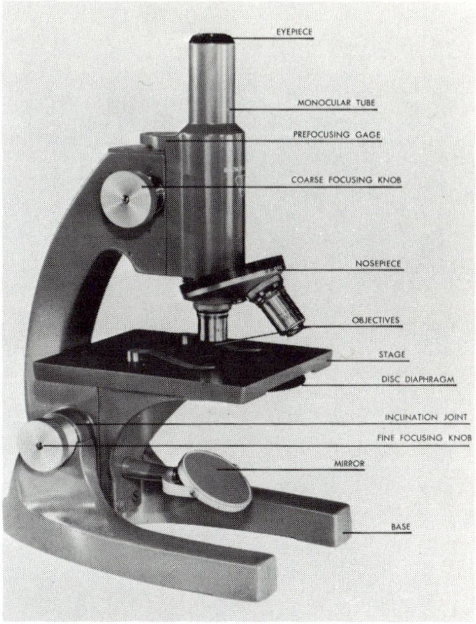

Fig. 83–1b

The modern miscroscope on the right is a big improvement over Robert Hooke's microscope shown on the left.

4. KNOWING THE PARTS OF YOUR MICROSCOPE

When a person learns to drive an automobile, he must first learn the names of the controls and what job each part does. To use a microscope correctly, you should know the parts of the microscope and their jobs. Let's look at Fig. 83–1b.

5. PREPARING THE OBJECT TO BE VIEWED

Objects to be viewed under a microscope must be very thin. The reason for this is that enough light must pass *through* the object and *through* the lenses to reach your eye. Objects which allow some

Fig. 83–2

How is this microscope different from the one in Fig. 83–1b? What are its advantages?

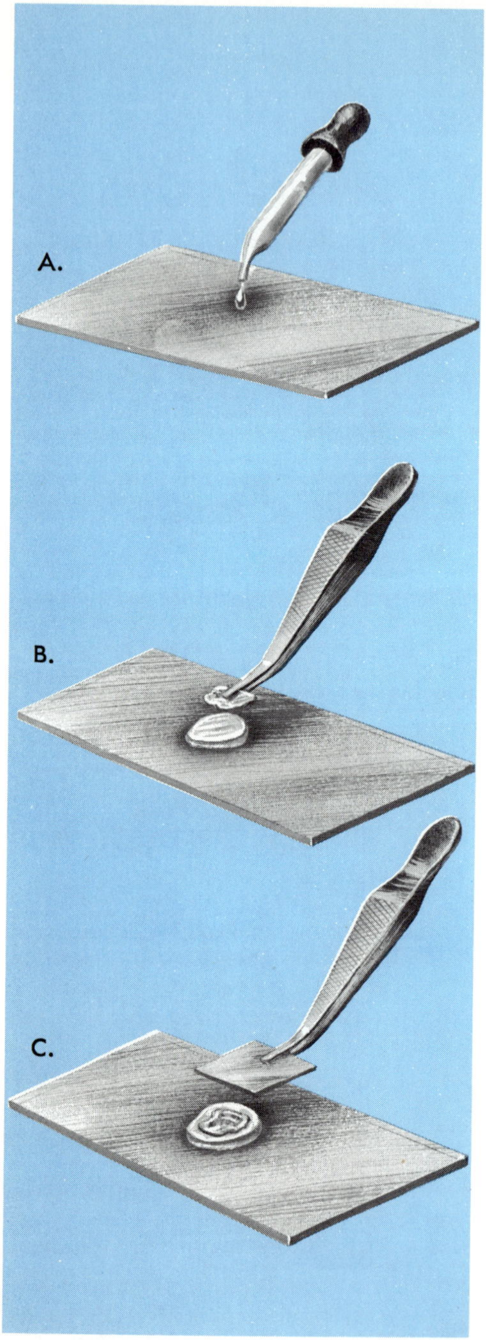

Fig. 84–1

Preparing a slide for the microscope. Explain each step.

light to go through them are said to be TRANSLUCENT (trans-LOO-sent). Frosted glass, soda pop and wax paper are examples of translucent materials. Parts of living things must, therefore, be peeled very thin. For example, you cannot view a whole leaf under the microscope. You must peel off a very thin layer. Objects to be viewed are placed on a thin glass slide, usually in a drop of water. The drop is then covered with a thin plastic cover and the slide is placed on the stage over the hole in its center.

CAUTION

- A microscope is a very expensive instrument.
- Your teacher will give you full directions on how to use it. Use it carefully!

6. LEARNING TO USE THE MICROSCOPE

After the slide is on the stage, you must lower the clips to hold the slide in place. You can swing either the low-power lens or the high-power lens into place. Either way, the objective lens must be in line with the eye lens. The mirror is placed in such a way that light passes through the hole and the slide and through the object. The two sets of adjustment wheels help you to FOCUS (FO-kus). This means "to make the image clear." You focus by raising the tube.

7. NOW, LET'S LOOK AT THE REAL THING . . .

The chart, below, shows how *you* can use the microscope to study tissue from your body. In the next few chapters, we shall "zero in" on what we can learn. Meantime, let's learn the *process* of studying cells.

YOU NOW KNOW

- Objects to be viewed with a microscope must be translucent.
- The mirror on a microscope directs light through the object being viewed.
- The knobs on a microscope help you focus for a clear image.

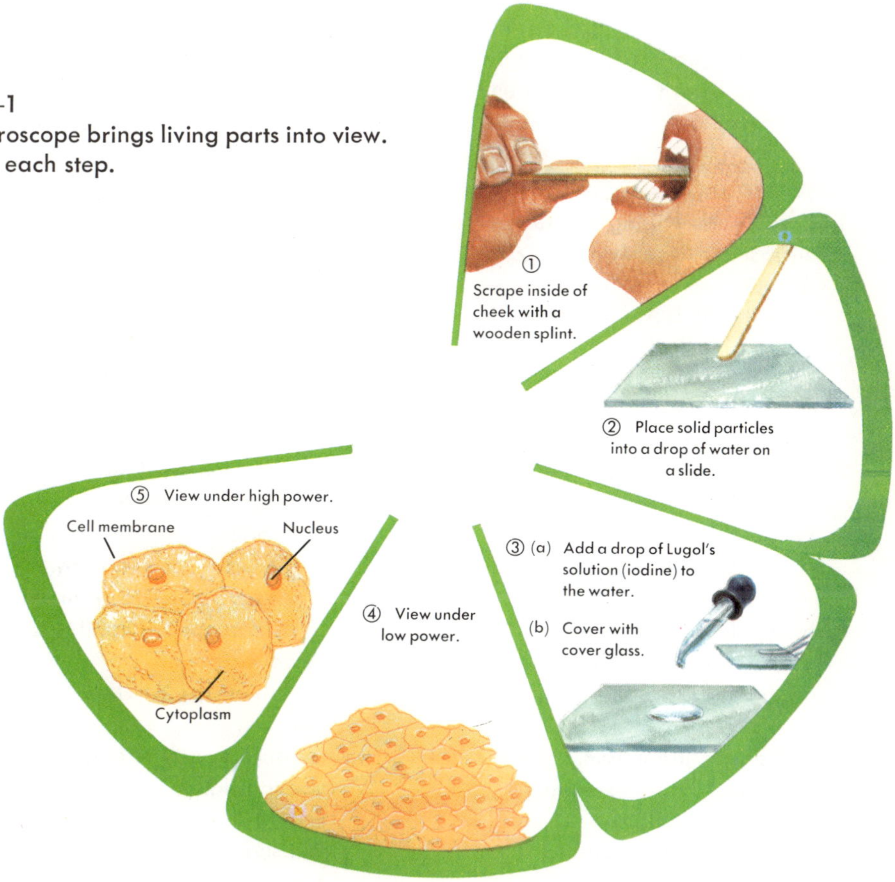

Fig. 85–1
The microscope brings living parts into view. Explain each step.

UNDERSTANDING WHAT YOU HAVE READ

I. What are the two main ideas of Chapter 13?

1. To focus is to make clear.
2. Microscopes use lenses.
3. A compound microscope has two adjustments for focusing an image.
4. Lenses magnify objects.

II. In which paragraph do you find the answer?

1. How many lenses are used in a compound microscope?
2. What are the names of the lenses of a microscope?
3. What is the purpose of the coarse adjustment on a microscope?
4. Why must an object be translucent in order for it to be viewed under a microscope?

III. Knowing What and Why

1. If you keep raising the tube of the microscope, you won't see anything because
   a. you will have no more light.
   b. the object will go out of focus.
   c. the mirror will no longer work.
   d. you will no longer be able to use the eye lens.
2. You cannot use all three lenses on the student microscope (Fig. 83–2) because
   a. you can line up only two of them at one time.
   b. one lens is too weak.
   c. the lenses don't match.
   d. the mirror does not work this way.
3. The simple magnifying glass would be *least* useful to examine
   a. bread mold.
   b. the inside of a plant's stem.
   c. a mosquito's eye.
   d. a bee's leg.
4. The most important difference between a cheap microscope and an expensive microscope is the difference in
   a. the focus-adjusting wheels.
   b. the stage and its controls.
   c. the metal and finish.
   d. the quality of the lenses.

IV. Complete the sentences.

1. Another name for the eye lens is the _____ lens.
2. The adjustment wheel is used for _____.
3. The mirror _____.
4. The base _____.
5. The clips _____.
6. Objects which let some light through them are _____.

a. translucent
b. reflects light
c. hold the slide
d. supports the microscope
e. ocular
f. focusing

V. Number, Please . . .

1. Suppose a microscope has an eye lens with a power of 10X and a low-power objective lens of 12X. How much does this microscope magnify under *low* power?
2. Suppose in the same microscope there is a high-power objective lens of 35X. How much does this microscope magnify under *high* power?

VI. Explorations for the Student Scientist

You can buy a good hand lens for about a dollar.

It will give you a great deal of pleasure to examine small objects with it. Examples: bread mold, insect, leaf, hair, rock sample.

14

TARGET
What does the microscope show us about living things?

The Microscope and the Unseen World

1. A NEW WORLD COMES INTO VIEW

TINY ANIMALS SEEN THROUGH MICROSCOPE

DUTCH SCIENTIST FINDS LIFE NEVER SEEN BEFORE

Special to the Holland Express
DELFT, HOLLAND, January, 1675—Letters to the Royal Society of London.

Fig. 87–1

This is an imaginary headline from an imaginary newspaper. It tells a story as it might have been told if it had happened today. Actually, it occurred a long time ago. Let us piece the story together as it really happened. A Dutchman, ANTON VAN LEEUWENHOEK (LAY-ven-hook), working as a janitor in the City Hall of Delft in Holland, had an interesting hobby. He ground lenses. Lenses have always been made from

Fig. 87–2
Anton van Leeuwenhoek, 1632–1723 He found a new world under the microscope.

glass by a very careful grinding process. With these lenses he examined everything he could find: drops of rainwater, tiny insects, saliva, decaying plants, yeast and hundreds of other objects and materials. Leeuwenhoek used these lenses singly—as a simple microscope. But they were very powerful lenses; that is, they could magnify things many times their true size. One day he was examining drops of water in which some plants had decayed. He was amazed to find hundreds of tiny creatures swimming back and forth. Never before had anyone imagined that such tiny living things existed! Never before had anyone reported seeing such creatures! A new world had come into view.

2. NATURE'S "LITTLE LEAGUE"

Leeuwenhoek made careful notes and drew diagrams of the things he saw. He called the little animals "beasties." He wrote of his discovery to English and French scientists. They checked his work and found that it was true.

What had Leeuwenhoek actually found? He found little animals which in their very tiny "bodies" could do all the life jobs you can. Such simple animals are called PROTOZOA (pro-to-ZO-uh) or "first animals." Later, other scientists found tiny plants which could be seen only with a microscope. The chart and diagrams at the end of this chapter will give you an idea of what some of these microscopic plants and animals are like. See pages 90–91.

3. THE BUILDING BLOCKS OF LIVING THINGS

At about the same time that Leeuwenhoek was studying his "beasties" (1665) an English scientist by the name of Robert Hooke was using a compound microscope for studying various parts of plants. One day, in examining a piece of cork, he noticed the regular pattern shown in the diagram below.

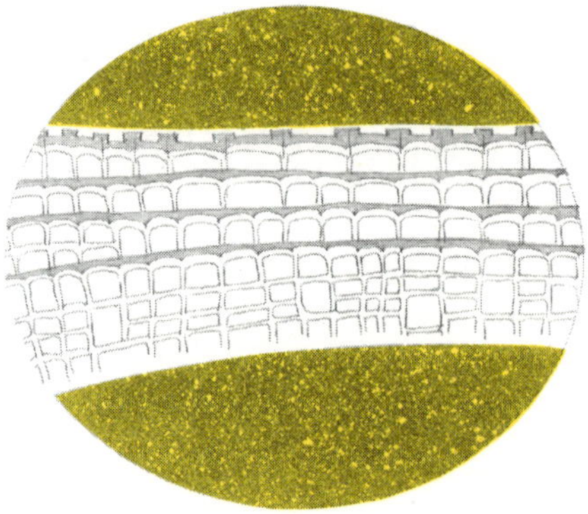

Fig. 88–1
Diagram of cork from Hooke's notebook.

This pattern reminded Hooke of the honeycomb of a bee which also shows this pattern of little rooms. In his book, he used the Latin word for "room," *cella*. Its English translation is CELL. Since the cork was part of a tree that was once alive, Hooke thought that all living things might be made of cells. He and other scientists continued to study the parts of living things. Always, they found that these living things contained cells.

4. THE CELL THEORY

By 1838, most scientists were convinced that all living things were made of cells. The CELL THEORY was then announced. A theory is a set of ideas which explains something. The most important ideas of the cell theory were:

- All living things are made of cells.
- All life activities of a living thing are really being done by its cells.

Look at the charts on pages 90–91. Yes, you are right! Each of these microscopic creatures consists of *only one* cell. Other plants and animals, including human beings, contain millions or even billions of cells. Today the cell theory is an accepted scientific idea.

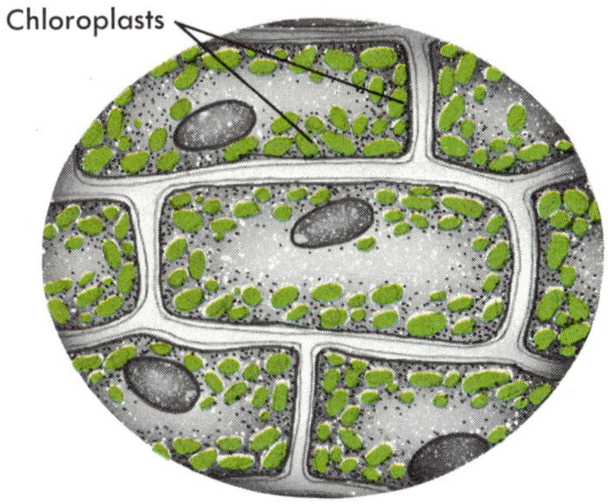

Fig. 89–1

Examples of plant cells. Are all plant cells green?

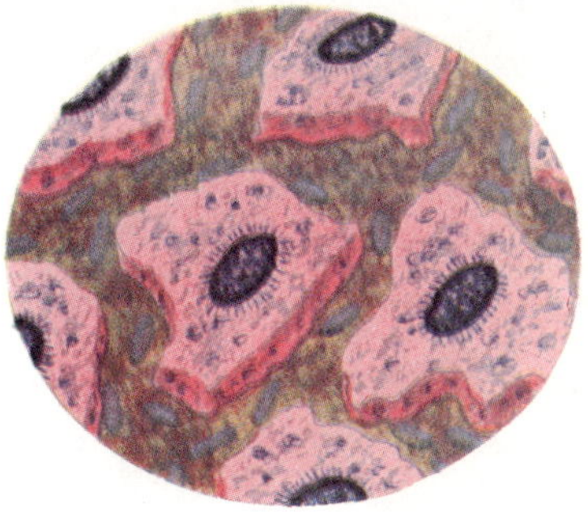

Fig. 89–2

Examples of human cells (inside of cheek). Is the color natural? Explain.

YOU NOW KNOW

- All living things are made of cells.
- Some simple plants and animals have only one cell.
- All life functions are done by one or more cells in the plant or animal.

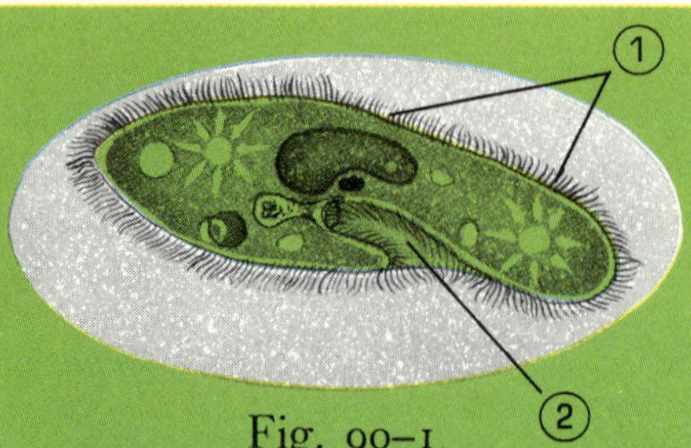

Fig. 90–1

PROTOZOA
Microscopic
animals

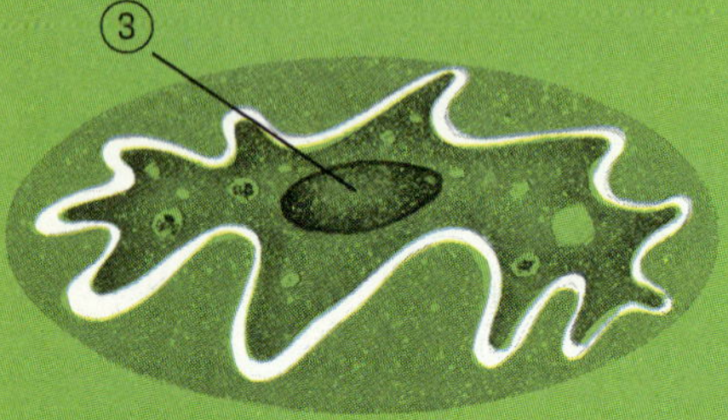

Fig. 90–2

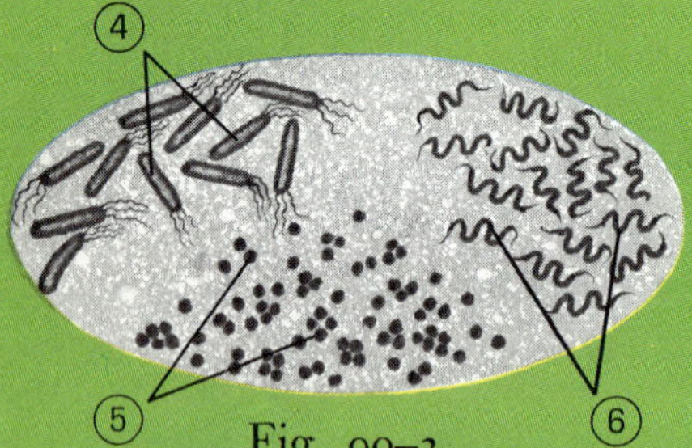

Fig. 90–3

FUNGI
(FUN-jy)
plants
without
green-plant
color

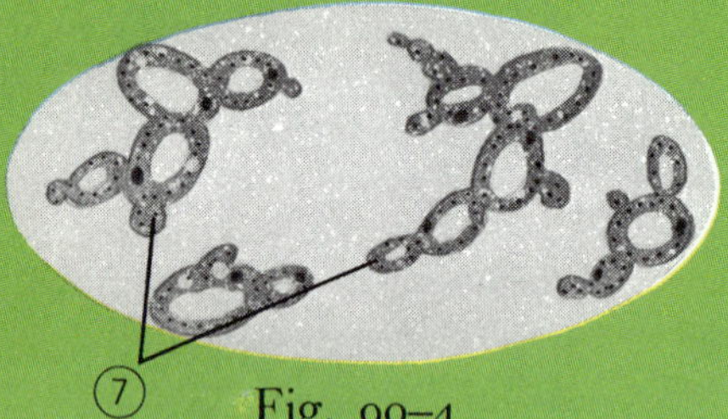

Fig. 90–4

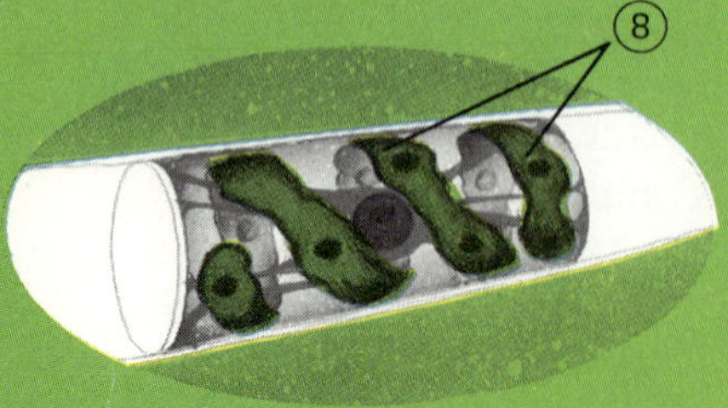

Fig. 90–5

ALGAE
(AL-jee)
plants
containing
green-plant
color

PARAMECIUM (par-a-MEE-see-um)
Animal with unchanging shape. Swims about by using hairs called CILIA (SIL-ee-uh) (No. 1).
No. 2 is the mouth.

AMEBA (a-MEE-buh)
This animal keeps changing its shape. Notice how it can surround food to digest it (No. 3).

BACTERIA (bak-TEE-ree-uh)
These simple colorless plants cannot make their own food. Some cause diseases. Some are rod-shaped (No. 4). Some are ball-shaped (No. 5) and some are shaped like corkscrews (No. 6).

YEAST (YEEST)
Simple colorless plants. They reproduce by growing buds (No. 7) which break off to form new plants.

SPIROGYRA (SPY-row-JY-ra)
This green plant lives in ponds. The spiral-shaped part (No. 8) contains CHLOROPHYLL (KLOR-uh-fil), the green chemical that these plants use to make their own food.

UNDERSTANDING WHAT YOU HAVE READ

I. What are the two main ideas of Chapter 14?

1. Leeuwenhoek discovered one-celled animals.
2. The microscope shows us that all living things are made of cells.
3. Some living things do not have chlorophyll.
4. All life functions are made possible by the work of the cells.

II. In which paragraph do you find the answer?

1. Who was the first scientist to describe cells?
2. What is the "Cell Theory"?
3. What are bacteria?
4. What are some kinds of one-celled plants?
5. How was the world of microscopic animals discovered?

III. Knowing What and Why

1. Leeuwenhoek used a simple microscope to discover
   a. the Cell Theory.
   b. microscopic life.
   c. cork cells.
   d. spirogyra.
2. The cilia of a paramecium
   a. move it through the water.
   b. cause it to reproduce.
   c. do nothing.
   d. help digestion.
3. Hooke saw and named cells in
   a. bacteria.
   b. ameba.
   c. yeast.
   d. cork.
4. Algae are plants containing chlorophyll. Therefore
   a. they are like bacteria.
   b. they are like yeast.
   c. they can manufacture their own food.
   d. they are like amebas.
5. Protozoa include:
   a. bacteria and paramecia.
   b. paramecia and amebas.
   c. yeast and bacteria.
   d. amebas, paramecia and bacteria.

IV. Match the Words

1. the man who thought of the term "beasties"
2. the man who named cells
3. fungi
4. microscopic green plants
5. yeast cells

a. Hooke
b. bacteria
c. buds
d. Leeuwenhoek
e. algae

V. For the Home Laboratory

You can grow yeast cells! Follow these directions:

1. Buy a small yeast cake in the supermarket.
2. Fill an eight-ounce jar with water. Add a tablespoon of sugar. Crumble half the yeast cake into the sugar solution. Cover and let the jar stand at kitchen temperature for two days.
3. Bring the jar to class to have its contents examined with a microscope.

15

TARGET
What can we see inside the cell?

A Look Inside the Cells

1. PUTTING THE EVIDENCE TOGETHER

Look at a brick wall from afar. It looks solid. Now come closer and look again. What do you see? You will probably notice that it is made of many bricks, each identical to one another. In the same way, all plants and animals are made of cells, the building blocks of all living things.

Many years have passed since the Cell Theory was announced in 1838. Since that time, as you can imagine, thousands of scientists have studied cells. Better microscopes have been built. Thousands of plants and animals have been *dissected*. As we learned earlier, to dissect is to cut up into smaller parts. These smaller parts could be examined with the new and better microscopes. Also many chemical tests have been made on cells. When we put all the facts together, we can be sure of three things:

1. In general, most cells have the same appearance.
2. There are chemical actions taking place in all cells.
3. These chemical actions explain the life functions.

2. WHAT DOES A CELL LOOK LIKE?

Although there are many kinds of dogs, you can easily describe in a general way what a dog—any dog—looks like. You can then go on to tell the difference between a shepherd and a collie. But, in general, all dogs look alike. The same is true of cells. Plant cells differ somewhat from animal cells. In plants, there are many kinds of cells; in animals, also, there are many kinds of cells. But we can still describe a *typical* cell, a cell which in general is like all cells. The two diagrams in Fig. 94–1 show typical plant and animal cells.

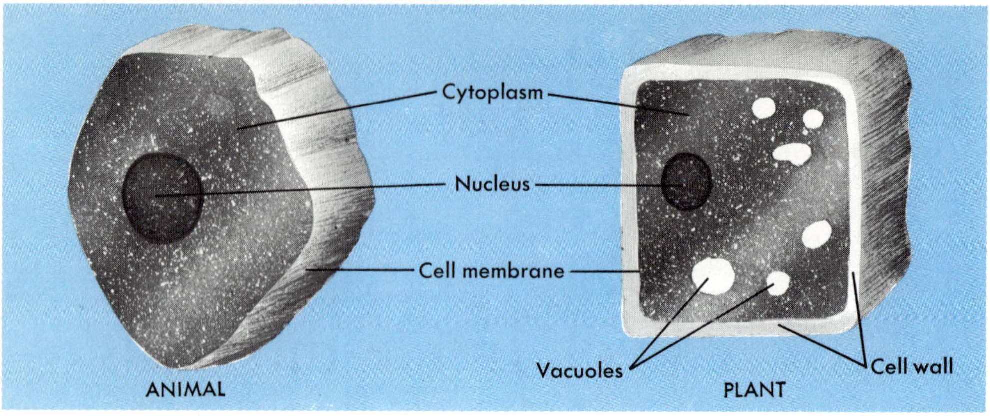

Fig. 94–1
Typical cells and their parts. Is it correct to show the plant cell without chloroplasts? Explain.

3. PLANT CELL OR ANIMAL CELL?

The cells Hooke saw seemed to have a boundary or wall. We call the outside of the cell the CELL WALL. Just inside the cell wall is another, thinner boundary called the CELL MEMBRANE. Compare the animal cell with the plant cell. Both cells have a cell membrane, but one chief difference between the two is that only the plant cell has a cell wall. In Fig. 94–1 a non-green plant cell is shown. Such cells are found in roots and stems. Other cells in leaves contain a green coloring. This green coloring comes from the chemical CHLOROPHYLL Chlorophyll is contained in small, ball-shaped objects in the cell called CHOROPLASTS (KLOR-uh-plasts). (See Fig. 89–1, which shows plant cells of this type.)

YOU NOW KNOW

- Plant cells differ from animal cells in that they have a cell wall outside of the cell membrane.
- Some plant cells have chloroplasts which contain chlorophyll. Such cells are green.
- Cells have length, width and thickness.

4. INSIDE THE CELL

Just inside the cell membrane is a thick fluid very much like egg white. This is the CYTOPLASM (SY-tow-plazm). Somewhere in the cytoplasm there appears a darker, ball-shaped object called the NUCLEUS (NEW-klee-us). Also inside the cytoplasm there are one or more "spaces" called VACUOLES (VAK-u-olz). Actually, these spaces are not empty. They are filled with water and dissolved chemicals. For a long while, living matter was called PROTOPLASM (PRO-tow-plazm). Today, we simply call it *living matter*.

When Hooke first saw cork cells, he saw only the cell walls, for the living matter had died. By using fresh material, we can see the living matter of plants and animals.

DO AND DISCOVER

Investigation 10: How can we show that green leaves contain chlorophyll? Procedure:

Heat several leaves which are both green and white in boiling hot water. (Example: silver-edged geranium) Then transfer the leaves to hot alcohol. Remove the leaves and spread them on a glass plate.

Experiment Analysis

1. Why are the leaves boiled first?
2. Why are we using hot plates and not open-flame burners?
3. Does light pass through the chlorophyll mixture in the last picture below?

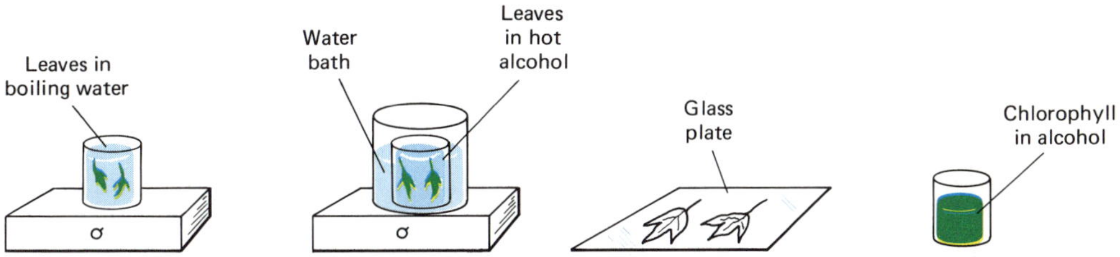

Fig. 95–1

LIVING MATTER
- Cytoplasm
- Nucleus
- Cell membrane
- Chloroplasts

5. SUMMARY OF CELL FACTS

Cell Shape	**can be**	Spherical—like a marble Rod—like a pencil Rectangular—like a brick Irregular—any shape
Cell Size	**can be**	Microscopic (2,500 cells laid end to end will measure one inch.) (Birds' eggs are exceptions; an ostrich egg, which is a single cell, is 8 inches in diameter.)
Color	**can be**	Colorless—like water or egg white Green—if the cell has chlorophyll Any color—depending on the color of the chemicals in the cytoplasm
Physical State	**can be**	Hard—as in bone or wood Soft—as in skin or leaves

UNDERSTANDING WHAT YOU HAVE READ

I. What are the two main ideas of Chapter 15?

1. All cells have similar parts so that most cells look alike.
2. Cytoplasm is different from nuclear material.
3. Vacuoles are definitely not empty.
4. The cell parts can be described when seen under a microscope.

II. In which paragraph do you find the answer?

1. What is a "typical" cell?
2. What is cytoplasm?
3. What is a chloroplast?
4. What does a vacuole look like?
5. What is the difference between a cell wall and a cell membrane?

III. Knowing What and Why

1. The color of cells is
   a. green.
   b. colorless.
   c. many colors.
   d. like egg white.
2. Living matter includes
   a. cytoplasm and a nucleus.
   b. a cell wall and a nucleus.
   c. a cell wall and a cell membrane.
   d. a vacuole and a cell wall.
3. A cell wall is found in the cells of
   a. bacteria only.
   b. all plants.
   c. only green plants.
   d. all living things.
4. The cells which Hooke first saw in cork consisted of
   a. a nucleus and cytoplasm.
   b. a nucleus and a vacuole.
   c. cell walls only.
   d. chloroplasts and a nucleus.

IV. Complete the story by giving the missing words.

Plant cells have a (1)____, but animal cells do not. All cells are made of a (2)______. A darker, ball-shaped object in the cell is called the (3)______. Some cells are colorless. The color of those with chloroplasts is (4)______.

V. For the Amateur Scientist

1. Using modeling clay or balsa wood, make a model of a typical cell. Bring your work to class.
2. Collect pictures of plants of different colors. Paste them on paper. Explain what gives these plants their colors. (Hint: beets, carrots, turnips, green beans, corn.)

16

TARGET
How does a cell do its work?

Getting the Work Done Inside a Cell

1. WHAT PART DOES WHAT?

By this time you have probably been wondering what the various parts of a cell do. You have probably guessed that each part has a different job. How can we find out what the job of each part is? There are two ways:

1. The shape, the form and the location of the part will give us some clue about its job.
2. The study of its chemical composition can tell us what job a cell part does.

2. THE CELL MEMBRANE FOR ENTRANCE AND EXIT

You have already learned the word *absorption.* Do you remember that it means "to take in"? Later, you learned the word *excretion.* Do you remember that it means "to send out"? If you look back to Unit II, Chapter 10, you will find an example

DO AND DISCOVER

Investigation 11: How do membranes control passing of materials into and out of cells?

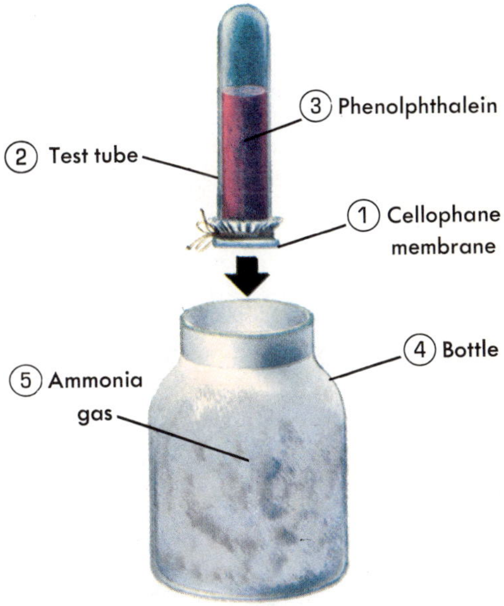

Ammonia gas or vapor (5) can cause phenolphthalein (3) to turn pink. Lower the test tube (2) into the bottle (4). In this experiment the pink color appears in the test tube. This proves that the gas can go through the membrane. Can you explain what this experiment has to do with what happens in a cell?

Fig. 97–1
Cellophane is like a cell membrane. How can gases be absorbed into cells?

of absorption and excretion. It is the life process known as respiration. In respiration, oxygen is absorbed and carbon dioxide and water are excreted. In the cell, absorption and excretion are controlled by the cell membrane. All things going in or out of a cell must pass through the cell membrane. The cell membrane is *not* solid. It has very small holes or pores which control what passes in and out of the cell.

DO AND DISCOVER

Investigation 12: Do dissolved materials pass through a cell membrane?

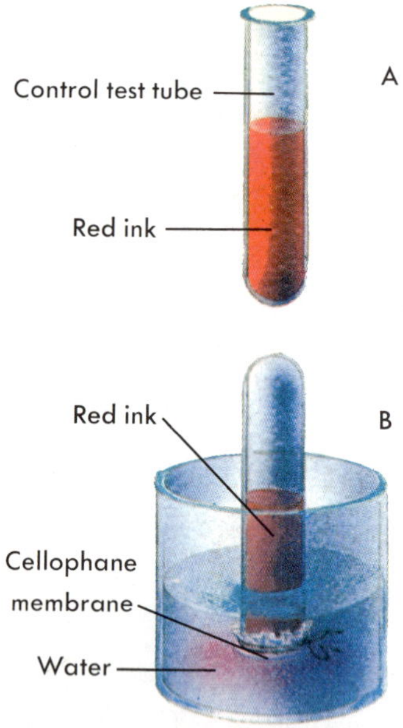

1. Can you describe this experiment in your own words?
2. Why does the ink in Test Tube B look pink after a few hours?
3. Why does the water in the glass jar turn pink?
4. What does this experiment prove?

Fig. 98–1

Why is Test Tube A called a "control"?

3. CELL WALLS FOR SUPPORT AND STRENGTH

Early scientists noticed that plant cells were generally more definite in their shape than animal cells. They soon realized that it was the thick cell walls that gave the plant cells support. Chemical tests showed that cell walls contain a compound called CELLULOSE (SEL-you-lohs) and that cellulose is *not* alive. As the cell grows, it manufactures the compound cellulose from food and stores it around the cell membrane. Now we can explain why trees are hard and strong. Trees have wood cells which have very thick cell walls with a great deal of cellulose in them. When the tree dies, the cell walls remain and we can cut the tree into lumber. Cotton fibers around cotton seeds also contain a great deal of cellulose. This is why cotton thread is strong. There is also a plant called flax which has strong cellulose fibers. These fibers are used to make linen cloth.

YOU NOW KNOW

- All cell membranes are alive. They control absorption and excretion.
- Cell walls, found only in plants, contain a nonliving material called cellulose.
- Cell walls give strength and support to plant cells.

4. THE NUCLEUS FOR GROWTH AND CONTROL

Scientists use tiny, delicate needles to dissect parts of a cell under a microscope. Slowly and very carefully, scientists have removed the nucleus. They have found that for a short time the cell continues to work. But soon the cell is no longer able to grow. It cannot reproduce. Chemical tests show that the nucleus is made of very complicated proteins. These proteins control assimilation (change of food into protoplasm), growth and reproduction. Within the cell are nucleic acids. The most important of these acids are DNA and RNA. DNA is found only in the nucleus. It "programs" the "blueprint" which makes offspring look like their parents. In other words, it controls *heredity*.

THE CHEMICAL MAKE-UP OF LIVING MATTER

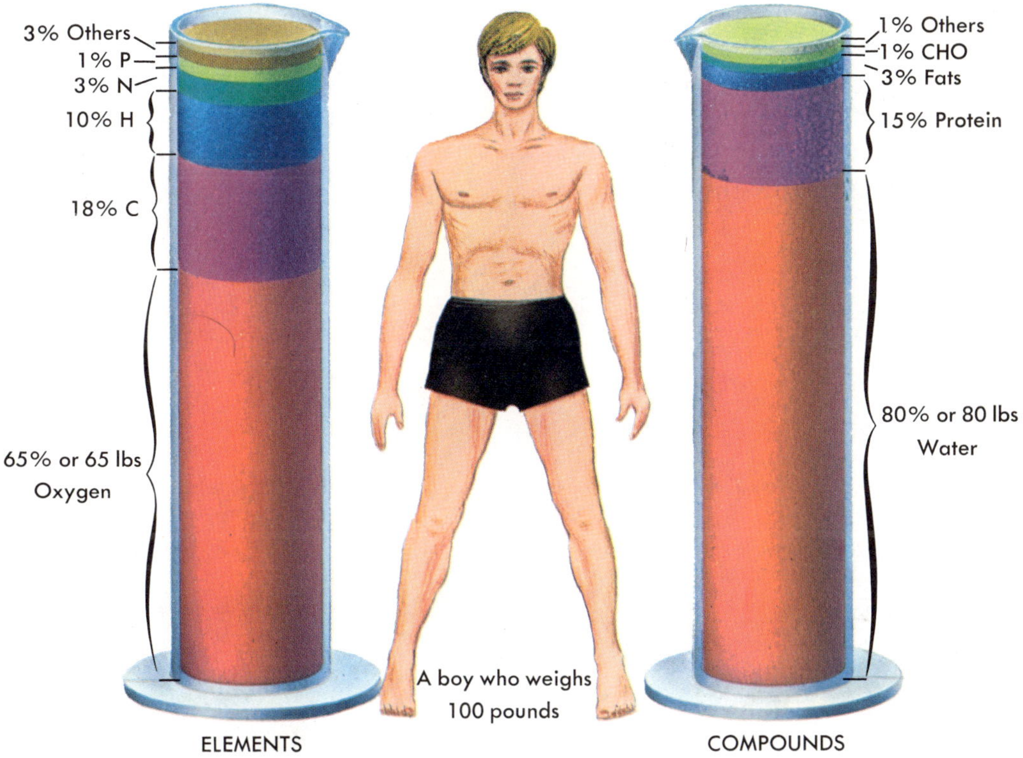

Fig. 99–1

Living matter contains ordinary elements and compounds in the approximate amounts shown above. How many pounds of water are there in this boy's body?

5. LIVING MATTER

Earlier, you read about alchemists who spent their entire lives trying to find some magic way to change cheap metals into gold. They never succeeded. An even deeper secret is the ability of living things to make protoplasm. Although we can now tell what chemicals are found in protoplasm, we cannot put these chemicals together to make this "magic life stuff." This means that all protoplasm now in existence has come from other living protoplasm. This colorless liquid is sometimes thin like water, sometimes thick like molasses. It is a mixture of ordinary elements and compounds. Protoplasm is the substance of life.

Fig. 99–1 shows the approximate amounts of elements and compounds which make up protoplasm. Think about these two important facts: first, protoplasm is a mixture of elements and compounds; second, there are chemical actions going on in cells all the time. That is why the composition of protoplasm keeps changing. For example, after digestion a cell may have more sugar. As respiration takes place, it may have more carbon dioxide and less oxygen.

YOU NOW KNOW

- Protoplasm is a complicated mixture of elements and compounds.
- As the life processes go on in cells, the composition of protoplasm changes.
- All life activities take place in protoplasm.

UNDERSTANDING WHAT YOU HAVE READ

I. What are the two main ideas of Chapter 16?

1. Protoplasm carries on life functions.
2. The parts of a cell have different jobs to do.
3. The cell wall is very different from the cell membrane.
4. The nucleus controls cell growth.

II. Explanation please . . . answer in one or two sentences

1. What part of a cell controls absorption?
2. Is protoplasm an element, a compound or a mixture?
3. Why do the contents of protoplasm keep changing?
4. Which part of a cell controls growth?
5. Which parts of a cell are not alive?

III. Knowing What and Why

1. Wood gets its strength from
   a. its bark.
   b. the water it stores.
   c. the cellulose in its cell walls.
   d. its protoplasm.
2. Cells are able to absorb
   a. liquids only.

b. liquids and gases.
c. gases only.
d. elements only.

3. The amount of sugar in a cell
   a. is always the same.
   b. is always changing.
   c. is not known to scientists.
   d. depends on whether it is plant or animal.
4. If the nucleus of an ameba is cut out
   a. the cell will not be able to reproduce.
   b. the cell will not be able to take in food.
   c. the cell membrane will not operate.
   d. excretion will not occur.

IV. Number, Please . . .

1. The most plentiful compound in protoplasm is _____, in the amount of _____ percent.
2. The most plentiful element in protoplasm is _____, in the amount of _____ percent.
3. The amount of nitrogen in protoplasm is _____ percent.
4. The amount of carbon in protoplasm is _____ percent.

V. For the Amateur Scientist—Do and Discover

Aim—To see whether salt, dissolved in water, can pass through a cell membrane.

Set up the experiment as shown below.

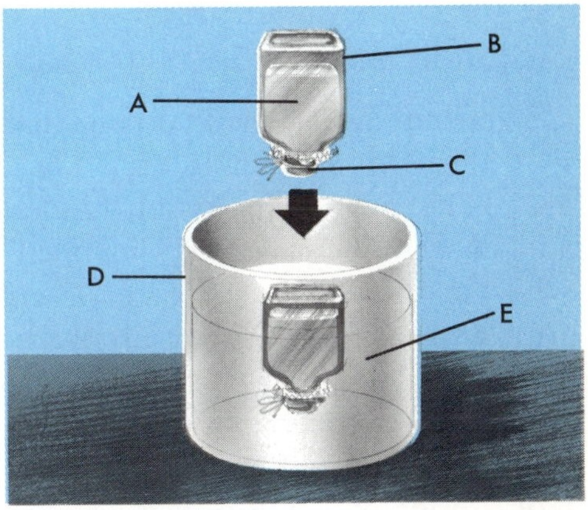

Step 1
Dissolve one teaspoonful of table salt in a jar of water.

Step 2
Place salty water (A) in a clean medicine bottle (B). Fix a piece of cellophane (C) around its mouth with a rubber band. Turn the medicine bottle upside down in a jar (D) partly filled with clear sink water (E).

Step 3
Allow to stand for 24 hours. Dip a wood splint into the water in the jar. Taste the splint. Is it salty? What have you learned?

Fig. 101–1

17

TARGET

How do groups of cells work together to perform life functions?

Teamwork Among Cells

1. ALL THE WORK OR A SPECIAL JOB?

In an automobile plant, no one man puts together the whole car. There are "body men" who attach fenders and doors. There are "spring men" who put the springs in place. There are "engine men," who install the engine, and so forth. The specialists add up to teamwork. In this chapter, we're going to talk about the teamwork inside of living things.

In Chapter 14 of this unit, we saw that some plants, such as bacteria and yeasts, and some animals, such as amebas and paramecia, consist of only *one* cell. Of course, all life functions in each of these cases are done by that one cell. But how do many-celled plants and many-celled animals function?

2. TISSUES FOR TEAMWORK

The microscope tells us the wonderful story of teamwork. We see that animals and plants are not just big masses of cells. We see that there are different *kinds* of cells and that cells form *groups*. The cells in these groups resemble each other in appearance. Of course, each cell must carry on *all* life functions. But because of their different structure, some cells are able to do one life task better than any other. When a cell concentrates on one life task we say that it is SPECIALIZED (SPESH-a-lized). Groups of cells that look alike and work together in a special life function are called TISSUES (TISH-uze).

3. COVERING TISSUES

Which of the tissues in your body is most visible to you? The tissue of your skin of course. And what do you think is its special job? It covers the cells below the surface. If you have ever fallen and scraped the top layer of skin, you were able to see the "raw" layers below. This is a clue to another job of covering tissue—it protects. The scientist's name for skin cells is EPITHELIAL (ep-i-THEE-lee-al) tissue.

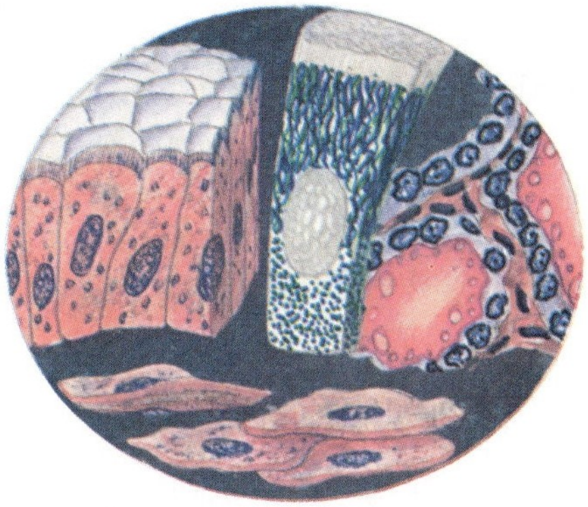

Fig. 103–1
Notice the different shapes of cells which make up epithelial tissue.

Epithelial tissue is found largely in the skin and in the linings of the body, where it covers the organs inside your body.

YOU NOW KNOW

- Cells work together as tissues.
- Epithelial tissue is one kind of tissue.
- It covers and protects.

4. TISSUES FOR SUPPORT

Most "higher" animals have a frame, or skeleton, to support the body. The skeleton consists of hard bone tissue. Under the microscope, it can be observed that living bone cells contain soft protoplasm. However, between the cells there are deposits of nonliving, hard compounds of calcium and phosphorus. It takes a hard blow to break a bone. Bones are even

DO AND DISCOVER

Investigation 13: What minerals give strength to bones?

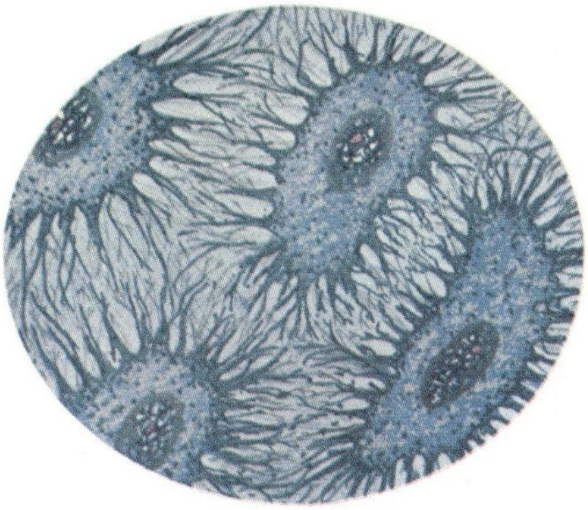

Leave a bone in a jar of vinegar for a few days. What happens? Explain.

Fig. 103–2
Bone cells store calcium and phosphorus compounds around them for strength. Vinegar is an acid which dissolves calcium. What can you now do to the bone? This is why a bone soaked in vinegar can be bent.

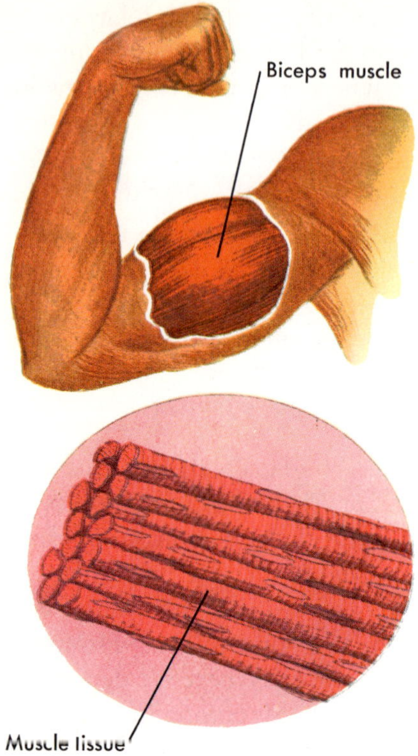

Fig. 104–1
Muscle tissue is used for locomotion and work.

strong enough to support the weight of an elephant!

Can you bend your ear? Can you bend the tip of your nose? Yes, but when you let go, the ear and nose spring back into place. The inner tissue of your ear is a flexible supporting tissue called CARTILAGE (KAR-til-ej). Cartilage is also found in the joints of bones, where it works as a shock absorber. For example, the knee joint has cartilage pads. When you jump, the shock is absorbed in your knees so that you feel the shock less in your spine, shoulders and head.

5. TISSUES TO MOVE AND WORK

Muscle tissue is made of cells which have no cell membranes between them and, therefore, appear as "bands" or "stripes." These bands grow longer or shorter to lengthen or shorten the muscle. Can you

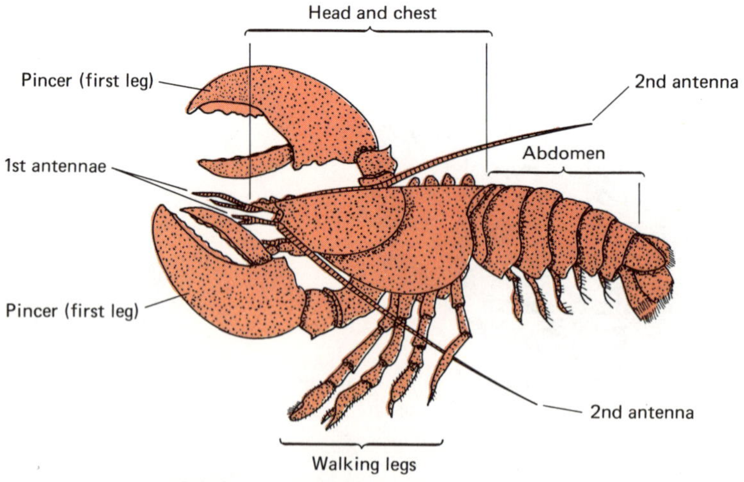

Fig. 104–2
The lobster has a skeleton on the outside! Explain the teamwork in this skeleton.

"make a muscle" in your arm? This muscle is called the BICEPS (BY-seps). Notice how you bend the arm at the elbow to make the biceps hard. You can choose to do this or choose not to do it. This muscle, like most muscles in your body, is controlled by your will. It is called a VOLUNTARY MUSCLE. There are some muscles in your body that are INVOLUNTARY—not under the control of your will. The muscles in the walls of your heart and in the walls of your stomach are examples of involuntary muscles.

YOU NOW KNOW

- Supporting tissue of bone and cartilage gives strength and support.
- Muscle tissue is for locomotion and work.

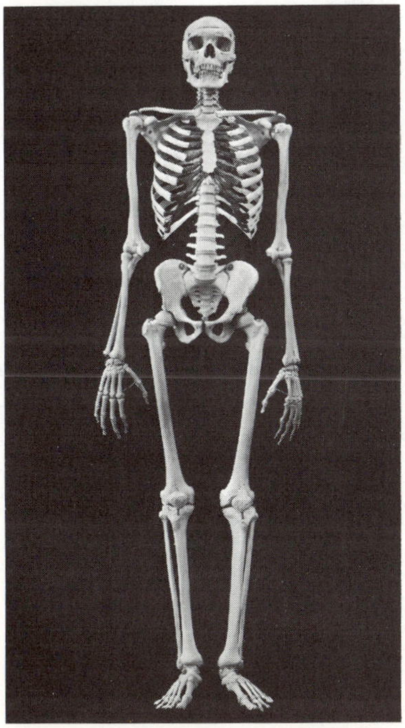

6. TISSUES TO CONNECT

Would you believe it? There are 206 bones in the human skeleton. How do they "hang together"? There is a strong connecting tissue which forms LIGAMENTS (LIG-uh-ments) that hold the bones together. There is also a connecting tissue which attaches bones to muscles so that the muscles have a firm hold for moving the bones. This connecting tissue is formed from cells that have fibers in them. Fibers which connect muscles to bones are called TENDONS (TEN-dunz). Strain or poor use of your body may cause you to stretch a ligament or a tendon. You will then have much pain and may be unable to move your arm or leg or shoulder. Doctors suggest rest, hot baths and massage to help the ligaments and tendons to heal.

Fig. 105–1.
Bones are held together by ligaments. There are cartilage pads between the bones. What is the advantage of the cartilage pads?

UNDERSTANDING WHAT YOU HAVE READ

I. What are the two main ideas of Chapter 17?

1. Groups of cells form tissues which have special jobs.
2. Tendons, ligaments, muscles and bones help you do your work.
3. All tissues are connected.
4. Tissues can do their special jobs because their cells have a special structure.

II. In which paragraph do you find the answer?

1. What is the scientist's name for skin cells?
2. What is the difference between a tendon and a ligament?
3. Why is bone hard?
4. What is the meaning of "tissue"?
5. How do muscles bend your leg at the knee?

III. Knowing What and Why

1. Tissues are formed from groups of
   a. similar cells doing the same job.
   b. similar cells doing different jobs.
   c. different cells doing different jobs.
   d. different cells doing the same job.
2. Tendons are *most* like
   a. cartilage cells.
   b. voluntary muscles.
   c. biceps.
   d. ligaments.
3. Milk, which is rich in calcium compounds, would be most helpful to
   a. build powerful muscles.
   b. develop a good skin.
   c. develop strong bones.
   d. give energy for heavy work.
4. Two special jobs of epithelial tissue are
   a. to cover and expand.
   b. to cover and protect.
   c. to protect and expand.
   d. to support and absorb shocks.

IV. Match the Words

1. specialized cells
2. phosphorus found in
3. tendons
4. voluntary muscle
5. biceps

a. connect bones to muscles
b. tissues
c. arm muscle
d. shoulder
e. bones

V. On the Ladder of Understanding

Choose the right ending.

1. Phosphorus and calcium are (elements or compounds).
2. The nucleus of a muscle cell controls (growth or absorption).
3. The outer boundary of a cartilage cell is the (cell wall or cell membrane).
4. The skin can receive a (stimulus or response).

18

TARGET
Which tissues command and control other tissues?

Tissues in Command and Control

1. BLOOD, THE LIFE-GIVING FLUID

A siren screams in the middle of the night. An ambulance races to deliver a bottle of precious liquid. It is PLASMA (PLAZ-muh)—the life-giving fluid in blood. Somewhere down the road, there has been an accident. The victim is "in shock." He lies pale and weak from loss of blood. Soon the ambulance arrives, and the ambulance doctor gives the man a TRANSFUSION (trans-FEW-zhun). After a short time, his color begins to return and he breathes more easily.

2. WHAT IS BLOOD?

In an average human body weighing 140 pounds, there are six quarts (or twelve pints) of a red liquid called blood. Blood is a tissue because it is made up of cells. Blood is different from other tissues, of course, because most of it is a liquid. You can examine a drop of blood under the microscope. You will easily see two kinds of cells floating in the liquid. There is also a third kind of cell that is harder to see.

YOU NOW KNOW

► Blood is a liquid tissue with cells floating in it.

3. THE LIQUID PART OF THE BLOOD

The liquid part of the blood is called plasma. It is a pale yellow liquid, mostly water. Look at Fig. 108–1. 92% of the plasma is water. The rest of the plasma contains other chemicals.

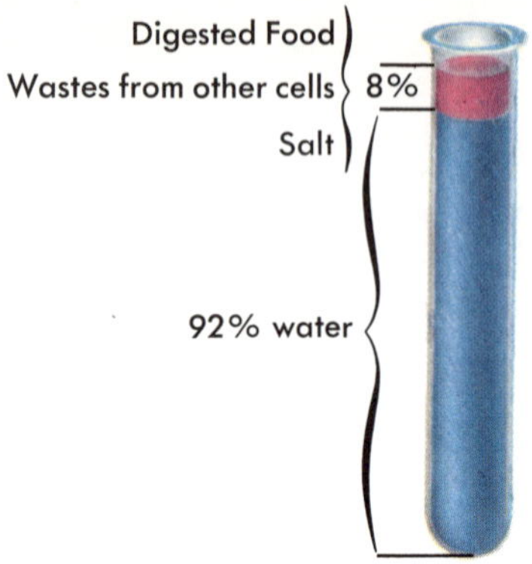

Fig. 108–1
The composition of plasma.

The amount of each of these other chemicals is always changing. A few hours after a good meal, there may be a great deal of sugar in your plasma. After vigorous exercise in the gym, your plasma contains a great deal of carbon dioxide. On a hot summer day, after you have perspired, there will be less salt in your plasma than on cooler days.

YOU NOW KNOW

- Plasma is the liquid part of the blood. It contains cells and dissolved chemicals.

4. BLOOD CELLS

A drop of blood under the microscope shows three kinds of cells. First, there are about five million RED BLOOD CELLS. These cells have no nuclei. Chemical tests show that they get their red color from a compound of iron called HEMOGLOBIN (HEEM-o-glo-bin). Second, there are about eight thousand WHITE BLOOD CELLS. These look like the ameba you studied earlier. Third, there are small, irregular cells called PLATELETS (PLATE-lets).

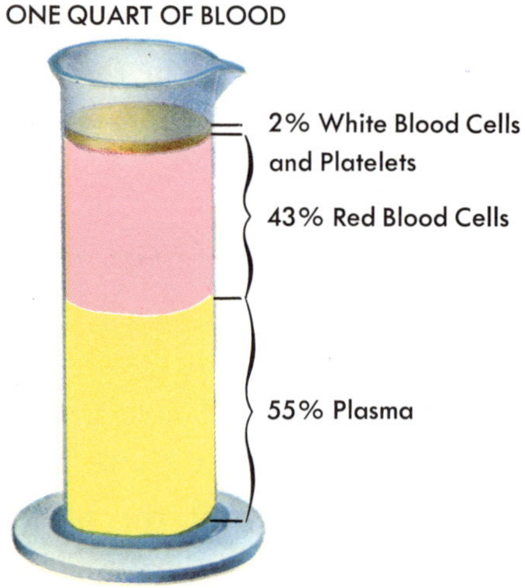

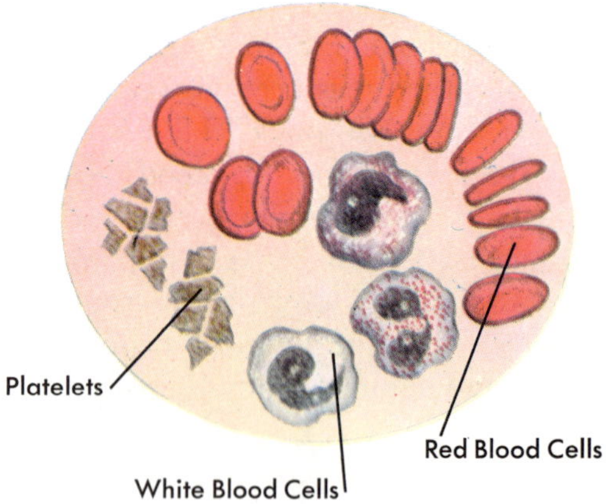

Fig. 108–2
Composition of blood.

5. WHAT IS THE WORK OF RED BLOOD CELLS?

Hemoglobin can form a temporary compound with oxygen. This temporary compound is called OXYHEMOGLOBIN. Oxygen joins with hemoglobin in the lungs, where air is breathed in. Blood carries the oxyhemoglobin to all cells. There the oxygen is absorbed by the cells. The red blood cells are now left with hemoglobin.

Red Blood Cells	In the Lungs: hemoglobin combines with oxygen.
	At the Cells: oxyhemoglobin gives off oxygen.

6. WHAT IS THE WORK OF THE OTHER BLOOD CELLS?

When you are healthy, a drop of your blood may contain about eight thousand white blood cells. When certain germs make you sick, however, there is a sudden increase in the number of white blood cells. Your "blood count" goes up. What happens is similar to our country drafting soldiers in time of war. When "enemy" germs invade our blood, our white blood cells go off to battle them. They surround and "swallow" the invaders. Sometimes white blood cells lose the battle, though. The germs destroy them and the result is PUS, the dead white blood cells and dead germs.

The platelets are much smaller than red blood cells and have irregular shapes. The platelets give off a chemical that causes the blood to clot at certain times. For example, when you cut yourself the blood clots at the wound to cut off the flow of any more blood. This saves you from losing a great deal of blood.

YOU NOW KNOW

- White blood cells destroy germs.
- Platelets begin clotting.

7. BLOOD CONTROLS MANY LIFE FUNCTIONS

Why does the title of Chapter 18 state that blood "commands" and "controls" other tissues? Let us remember what we read before: blood carries digested food to all cells. Blood carries oxygen to all the cells. This oxygen produces energy through the oxidation of food. Blood fights disease all over the body. No wonder blood has been called the "life-giving" liquid!

8. WHEN YOU LOSE BLOOD

When a patient has lost a lot of blood, new blood can be put into his body. The transfer of blood is called a transfusion. A transfusion can be made directly through a tube from one person to another. More often, the patient receives plasma or whole blood from a laboratory which is

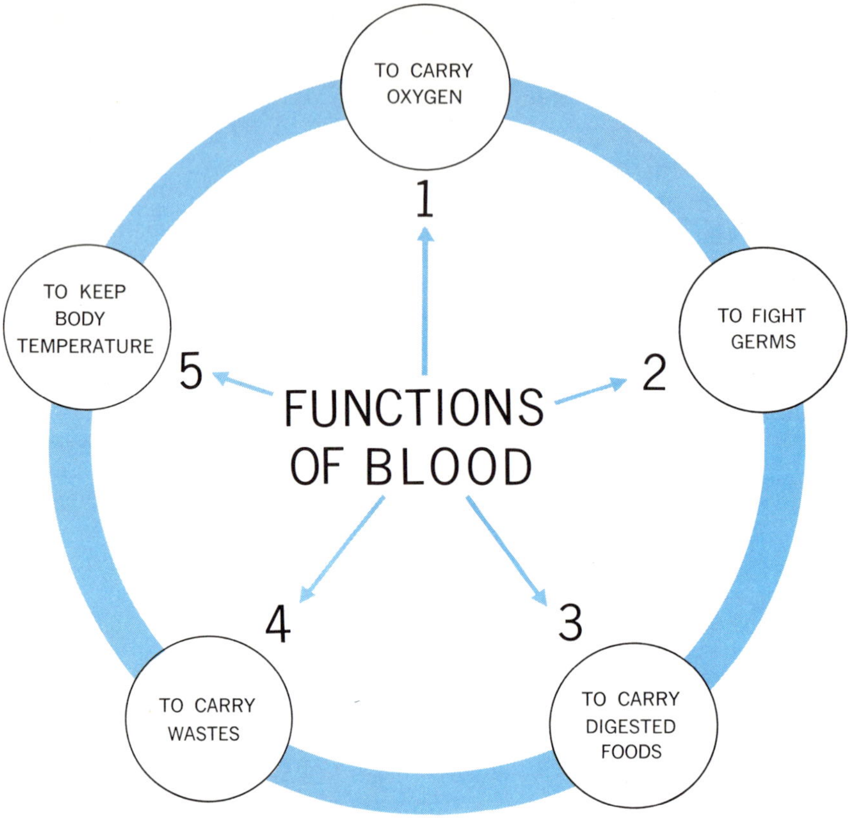

Fig. 110–1
Functions of blood.

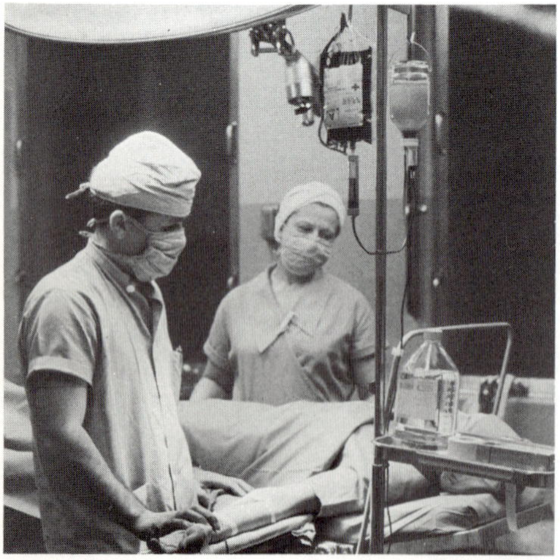

Fig. 110–2
A blood transfusion can save a life.

called a BLOOD BANK. Blood banks store large amounts of blood for emergencies.

9. NERVE TISSUE FOR COMMUNICATION AND CONTROL

Earlier, we spoke of stimulus and response. We gave the example of your finger touching a hot stove and pulling away very quickly. To pull your arm away, muscles had to be "told" to pull the arm bones. In other words, a message had to be sent, a command had to be given, and an "order" had to be carried out. How was this done? When your finger touched the hot stove, nerve cells in the tip of your finger received a stimulus. The nerve cells instantly sent a message to your brain. The brain then "read" the message and decided that the stove was too hot to touch. A command was then sent from your brain to the muscles in your arm. The brain "ordered" the muscles to pull back your arm and your arm responded. All this took place in an instant, too quickly for you to know what was happening. Your body contains many nerve cells which make it possible for you to feel, see, hear, taste, and smell the world around you.

10. THE NERVOUS SYSTEM

All these nerve cells are in some way connected to your brain. This system of nerves is called the NERVOUS SYSTEM. The function of the nervous system is to carry stimuli to the brain so that the body can respond to the outside world. The nerve cell shown in Fig. 111–1 has long branches which connect with thousands of other nerve cells. These cells are called NEURONS (NEW-ronz). They are extremely sensitive. Sometimes, we respond without realizing what we are doing. In the example given above, your arm pulled away without your will and without your thought. This is called an INVOLUNTARY RESPONSE. Before a baseball player at bat swings at the pitched ball, he also uses nerve cells. However, his response is under the control of his brain and will. He may decide to swing or let the ball pass. This is called a VOLUNTARY RESPONSE.

Animal Senses

SENSE	ORGAN
Sight	Eyes
Smell	Nose
Hearing	Ears
Taste	Tongue
Touch	Skin

YOU NOW KNOW

- Senses pick up stimuli.
- Your nerve cells send messages to the brain, muscles, bones and glands.

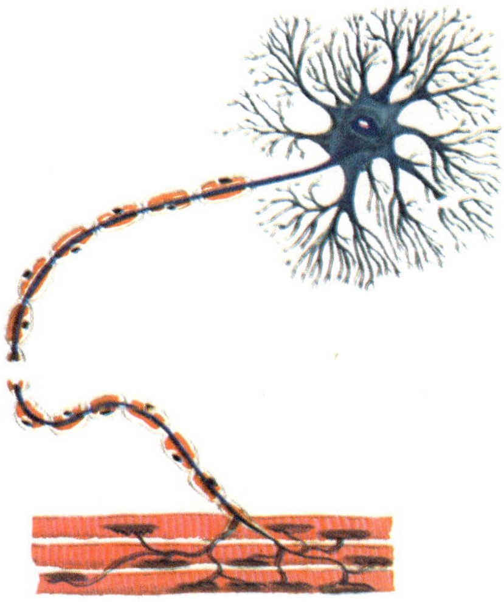

Fig. 111–1

Why have these tissues been likened to a telephone system?

UNDERSTANDING WHAT YOU HAVE READ

I. What are the two main ideas of Chapter 18?

1. All our energy comes from blood.
2. Transfusions stop bleeding.
3. Blood is a tissue.
4. When you are sick, your blood count goes up.
5. Red blood cells are more important than white blood cells.

II. Explanation please . . . answer in one or two sentences.

1. What senses are for protection?
2. How can we treat shock?
3. How is a clot formed?
4. How is pus formed?
5. Why do we usually use plasma or whole blood for transfusions?

III. Knowing What and Why

1. The number of white blood cells is greater when your tonsils are infected because white blood cells
   a. provide oxygen to give you energy.
   b. are needed to destroy germs.
   c. are found mostly in the region of the throat.
   d. stop shock.
2. Plasma is carried by Army medical men in powdered form because
   a. it is lighter.
   b. it is cheaper.
   c. it can give more energy.
   d. it can be taken by mouth.
3. People who have lost a great deal of blood are fed foods rich in iron because
   a. iron is needed to increase plasma.
   b. iron helps clotting.
   c. iron gives energy.
   d. iron is needed to form hemoglobin.
4. Since a red blood cell does not contain a nucleus, it cannot
   a. reproduce.
   b. move.
   c. use oxygen.
   d. give off carbon dioxide.

IV. What Number? What Size?

1. Which blood cells are most numerous?
2. Which blood cells are the largest?
3. Which blood cell has the largest nucleus?
4. What forms more than half of blood?
5. What forms more than 90% of plasma?
6. How many pints of blood does an average person have?
7. "Eight thousand cells per drop" probably refers to which blood cells?
8. Which are the *longest* cells in the body?

V. True or False

If the statement is true, write "true." If the statement is false, supply a word for the underlined word to make the statement true.

1. Sugar is found in <u>plasma</u>.
2. White blood cells "swallow" <u>germs</u>.
3. <u>Neurons</u> control clotting.
4. Nerve cells control <u>behavior</u>.

VI. On the Ladder of Understanding

1. Two elements found in hemoglobin are _____ and _____.
2. NaCl is the formula for _____.
3. The outer boundary of the white blood cell is called _____.
4. 92% of plasma has the formula _____.

19

TARGET
How do tissues combine to form the total plant or animal?

Teamwork Among Tissues

1. WHERE ARE TISSUES FOUND?

To answer this question, let us pick one small part of your body—your finger. The first thing you will probably notice is that your finger has a covering tissue. So you know that your finger has epithelial tissue. Now feel your finger for the bones. There are several bones and they are connected to each other. So your finger must have bone tissue and connective tissue. Can you move your finger? It is the muscle tissue which moves it. When you cut your finger, you feel pain with your nerve tissue. And the blood, flowing out of your cut, leaves no doubt that your finger contains blood tissue. And so the answer to the question above is: almost every part of your body has a little of each of the tissues.

2. TISSUES INTO ORGANS

Name a few parts of your body that do special jobs. The chances are that the part you name is an ORGAN. Here are examples of organs:

heart—brain—lung—stomach—
eye—kidney—skin

Each of these parts is made up of at least one tissue and usually more than one. For example, since the heart pumps the blood, it is mostly muscle tissue. But it also has connective fibers in its valves and epithelial tissue to cover it. The brain is your thinking and control organ. So it is mostly nerve tissue. But it also has blood in it.

YOU NOW KNOW

- An organ is one part of the body that has a special job.
- An organ has one or more tissues in it.

3. ORGANS INTO SYSTEMS

Look at Fig. 56–1 which shows you the parts of the body dealing with digestion. How many organs can you find? There is the mouth, liver, pancreas (PAN-kree-us), stomach, large intestine and small intestine. In paragraph 2 we saw that each organ does a special job. Now we see a new team; several organs that work together form a SYSTEM (SIS-tum). In this case, we are talking about the DIGESTIVE SYSTEM. Study the charts below for examples of other systems.

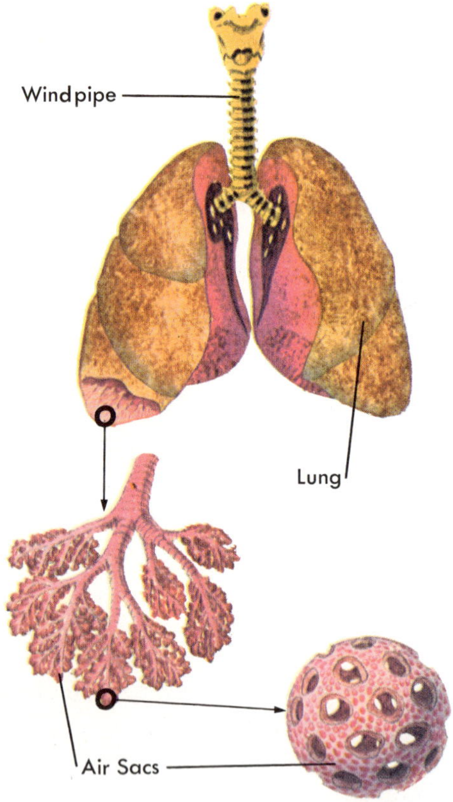

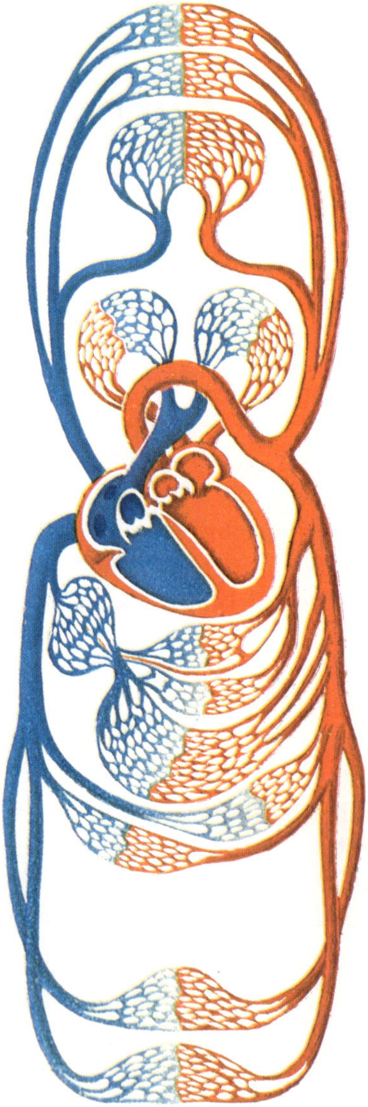

Fig. 114–1
How many organs can you find in each system? What are the two systems shown?

4. THE TISSUES OF PLANTS

So far we have spoken of tissues, organs and systems in animals. There are also plant tissues. Look at the diagram in Fig. 115–1.

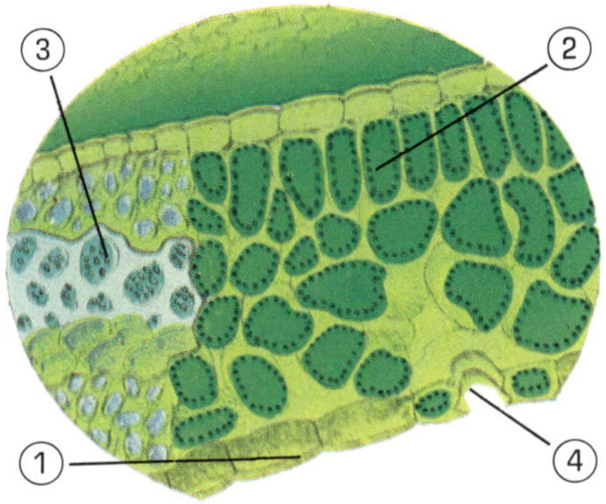

Fig. 115–1

A microscopic view through a typical leaf. What is the function of each part?

Notice that the cells on the outside of the leaf (No. 1) are all similar. You can guess that these cells, which are called EPIDERMIS (ep-i-DER-mis), are like epithelial tissue. The epidermis covers the top and bottom of the leaf and protects the inside cells. Now notice that the inside cells (No. 2) are green. They contain chloroplasts and chlorophyll. As you learned before, this tissue can make food for the plant. The SPONGY CELLS (No. 3) form a tissue with large spaces to store oxygen and carbon dioxide. These gases move in and out of leaves through the openings called STOMATES (STOW-mates) (No. 4).

Stems and roots in plants have special tissues to carry water and dissolved foods.

5. PLANT ORGANS

The major organs in most plants that you know are the roots, stems and leaves. Some plants also have fruits and flowers. As in animals, several organs work together to form systems. In plants, the most important systems are the food-making system, the water-carrying system and the reproduction system. Study the diagram below.

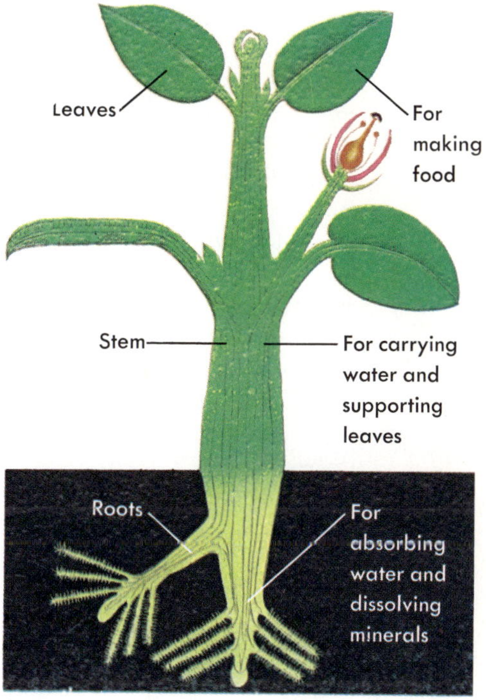

Fig. 115–2

The major parts of a typical plant.

UNDERSTANDING WHAT YOU HAVE READ

I. What are the two main ideas of Chapter 19?

1. Tissues combine to do similar jobs.
2. Epidermis is similar to epithelium.
3. Tissues are found all over the body.
4. Plants and animals have their parts organized to work together.

II. In which paragraph do you find the answer?

1. What part of a plant makes food?
2. What are the organs of the digestive system?
3. What are the parts of a typical leaf?
4. How are organs formed?
5. How do gases go in and out of leaves?

III. Knowing What and Why

1. The layer of cells on the outside of the leaf is called
   a. stomates.
   b. spongy cells.
   c. epidermis.
   d. epithelium.
2. The digestive system contains the following organs:
   a. lungs, stomach and mouth
   b. windpipe, gullet and stomach
   c. liver, small intestine and pancreas
   d. large intestine, small intestine and windpipe
3. The leaves of a plant get water
   a. from the soil through roots and stems.
   b. from the soil through stems only.
   c. from the air through stomates.
   d. from the air through epidermis.
4. The brain is mostly
   a. bone tissue.
   b. connecting tissue.
   c. nerve tissue.
   d. all tissues.

IV. Complete the story.

Use the words below.

Cells which are alike and do a similar job are called (1) ________. When we study a system, we find it is made up of (2) ________. The respiration system deals with (3) ________. A platelet is a (4) ________.

breathing
organs
tissues
excreting
muscle
nerve
cell

V. Put on Your Thinking Cap.

1. Why will smearing vaseline on the under-surface of a green plant's leaves kill the plant?
2. Does an organ contain more than one tissue? Explain.
3. Name three tissues found in your windpipe. What does each one do to make the windpipe work properly?

Glossary: What Does It Mean?

algae	AL jee	Microscopic one-celled green plants.
amebas	a MEE buhz	Microscopic one-celled animals that have a changing shape.
bacteria	bak TEE ree uh	Microscopic one-celled colorless plants which cannot make their own food.
biceps	BY seps	Arm muscle.
blood	BLUD	A liquid tissue with cells floating in it.
blood bank		A place for storing large amounts of blood or plasma for future use.
cartilage	KAR til ij	Flexible, supporting tissue.
cell	SEL	The unit of structure of all living things.
cell membrane	SEL MEM brayn	Inner boundary of a cell just inside the cell wall.
Cell Theory		The idea that all living things are made of cells which perform all the life functions.
cellulose	SEL you lohs	A nonliving compound contained in cell walls.
cell wall		The outside, or boundary, of a plant cell.
chlorophyll	KLOR oh fil	Green chemical compound found in plants which make their own food.
chloroplasts	KLOR oh plasts	Plant cell parts which contain chlorophyll.

cilia	SIL ee uh	Hairs on a paramecium cell used for locomotion.
compound microscope	KOM pownd	A microscope which combines two lenses (ocular and objective).
cytoplasm	SY tow plazm	Thick fluid material inside the cell membrane.
epidermis	ep i DER mis	Outside layer of cells in plants.
epithelial tissue	ep i THEE lee al	One kind of tissue which covers and protects.
focus	FO kus	To make an image clear.
hemoglobin	HEEM o glo bin	A compound of iron found in red blood cells.
high-power lens		The stronger of the two moving lenses in a compound microscope.
Hooke, Robert		English scientist who was the first to use a compound microscope to study plants.
involuntary muscle		A muscle not under the control of the will. Example, heart muscle.
Leeuwenhoek, Anton van	LAY ven hook	Dutchman who developed the microscope and was first to discover one-celled animals.
lens	LENZ	A specially prepared piece of glass that can magnify small objects.
ligaments	LIG uh ments	Connecting tissue which holds the bones together.
low-power lens		The weaker of the two moving lenses in a compound microscope.
microscope	MY kruh skope	An instrument used to see small objects.
neurons	NEW ronz	Nerve cells.
nucleus	NEW klee us	A ball-shaped object inside the cytoplasm that controls cell functions.
organ	OR gin	One part of a plant or animal which has a special job.
pancreas	PAN kree us	Organ of the digestive system which makes a digestive fluid.
paramecium	par a MEE see um	One-celled animal with unchanging shape.
plasma	PLAZ muh	The liquid part of blood, containing cells and dissolved chemicals.

platelets	PLATE lets	Cells in the blood which begin the clotting process.
protoplasm	PRO tow plazm	An older term used to describe all living matter.
protozoa	pro to ZO uh	Simple, one-celled animals.
pus		White blood cells which have been killed by germs, and dead germs.
red blood cells		Blood cells containing hemoglobin.
specialized	SPESH a lized	Able to do one task better than all others.
spirogyra	SPY row JY ra	A green plant which lives in ponds.
spongy cells	SPUN jee	Tissue in plants which stores oxygen and carbon dioxide.
system	SIS tum	Several organs working together.
tendons	TEN dunz	Fibers which connect muscles to bones.
theory	THEE uh ree	A set of ideas which explains something.
tissues	TISH uze	Groups of cells that are alike and work together in a special function.
transfusion	trans FEW zhun	The transfer of blood from one body to another.
translucent	tranz LOO sent	Allowing some light to go through.
vacuoles	VAK u olz	Spaces inside cytoplasm containing water and dissolved chemicals.
voluntary muscle		A muscle controlled by the will. Example: the biceps.
white blood cells		Cells in blood which fight germs.
yeast	YEEST	Colorless plants which reproduce by growing buds.

UNIT IV

Food provides energy for life. The green tree at the top and the plants on the left side use energy from the sun to make food. Animals use the food from the green plants. The food contains chemicals which enter the body of the man. Chemical changes in the body cause growth and provide energy.

FOOD—
THE
FUEL
OF
LIFE

UNIT IV

Food–The Fuel of Life

CHAPTER 20. SOME PLANTS MAKE THEIR OWN FOOD

CHAPTER 21. CHEMICAL CHANGE IN FOOD RELEASES ENERGY

CHAPTER 22. THE SEARCH FOR FOOD AND NUTRIENTS

CHAPTER 23. IDENTIFYING THE NUTRIENTS

CHAPTER 24. EAT RIGHT . . . LIVE RIGHT

CHAPTER 25. COUNTING CALORIES

CHAPTER 26. CONTROLLING YOUR DIET

20

TARGET
How do green plants make their own food?

Some Plants Make Their Own Food

1. IN THE SUMMER, WHEN LEAVES ARE GREEN . . .

In the summer, there is green everywhere: green-leaf trees, green-leaf bushes, green-leaf vines. And while the farms and orchards are green, there is fresh farm food. In the fall, there are ripe red apples, fat yellow beans, juicy melons, sweet berries and plenty of tomatoes. When the leaves fall, farm products are no longer fresh. What is the connection between green leaves and food?

2. LOOKING BACK

Turn back to page 52, Unit II, Chapter 8. Read paragraph 3 again. Look at Fig. 52–1.

YOU HAVE NOW REVIEWED

- Plants absorb water and minerals from the soil through their roots.

Now turn back to page 94, Unit III, Chapter 15. Read paragraph 3 again. Look at Fig. 89–1.

YOU HAVE NOW REVIEWED

- The cells of green plants contain chloroplasts which hold the green chemical chlorophyll.

3. GREEN PLANT CELLS ARE FOOD FACTORIES

Water is absorbed from the soil by the roots of plants. The water then

Raw Materials			Light Energy	Useful Food		Waste Gas
Water from Soil	+	Carbon Dioxide from Air	Chemical Change Caused by the Sun	Simple Sugar	+	Oxygen

$$6H_2O + 6CO_2 \longrightarrow C_6H_{12}O_6 + 6O_2$$

moves up into the leaves through the water tubes. In the meantime, the leaves have absorbed air through the stomates. The air contains the gas carbon dioxide. The green cells now have the raw materials for making food. Food-making is a chemical process which is very hard to understand. But the simple formula shown above will help you understand it.

4. ONLY IN GREEN PLANTS

We have said before that this food-making can be done only in green plant cells. Experiments show us that plants such as bacteria, molds and yeast (which are not green) cannot make their own food. They must get their food from other living things.

YOU NOW KNOW

- Chlorophyll, a chemical found in green plant cells, is needed for a plant to make its own food.

5. ONLY WITH LIGHT

Take a healthy, growing green plant and place it in a dark room. Even if you water it and keep the temperature right, it will soon die. Gardeners and farmers know that plants need sunlight to grow. The sun or some artificial light gives the green cells the energy needed to make food. The making of food by green plants is called PHOTOSYNTHESIS (fo-to-SIN-thuh-sis). Look at the experiment shown below. Can you explain what is happening?

DO AND DISCOVER

Investigation 14: What gas do green leaves form during food making?

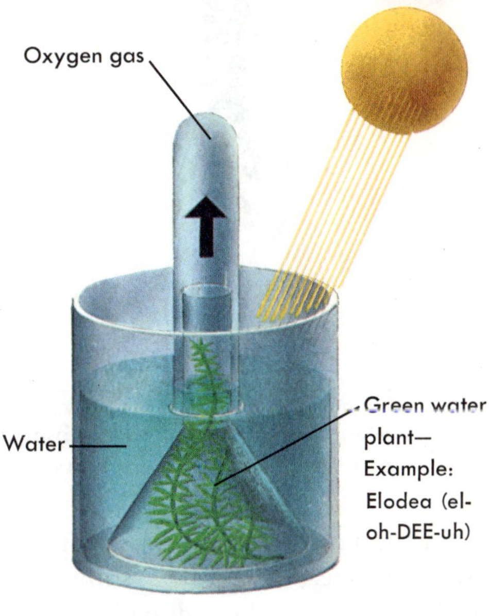

Fig. 125–1
Why will oxygen be formed only in sunlight? What is the test for oxygen?

photo	+	synthesis	=	photosynthesis
light	+	to combine	=	chemicals are combined to make food

DO AND DISCOVER

Investigation 15: What gas do green leaves absorb during food making?

Fig. 126–1
What gas did the elodea absorb? What process is the elodea carrying on? Will this change happen under artificial light?

In the investigation above, a green plant, Elodea, is used to show that plants absorb carbon dioxide during photosynthesis.

6. FROM SOIL TO FOOD

Green plants make food for themselves by photosynthesis. It is only natural that man and other animals should use plants and their parts for their own food. The chart on the following page shows examples of foods which we get from plants.

Plant Parts Store Food

Roots	Carrots Beets Turnips Sweet potato	Fruits	Melon Apples Plums Tomatoes Grapes
Stems	Celery Sugar cane (sap) Maple sugar (sap)	Seeds	Barley Rice Beans Wheat Corn Nuts
Leaves	Lettuce Cabbage Spinach		

Fig. 127–1

Food from the farm: Green leaves made these products through photosynthesis. Why is a tomato a fruit? Why is a cucumber a fruit? Is corn-on-the-cob a fruit? Explain.

YOU NOW KNOW

- Through photosynthesis, green plants produce food.
- Plants store food in their parts.
- Animals use plants for food.

7. "GREEN WATER AHEAD!"

"Green water ahead!" is a cry often heard on a deep-sea fishing boat. It means that the water ahead is filled with millions of microscopic green plants. Green means chlorophyll; chlorophyll means food—food for fish. No wonder fishermen know that blue water means poor catching; green water means good catching.

8. GARDENS OF THE SEA

Does the title surprise you? It should not. Yes, the sea is the home of billions of plants. There are no big trees or shrubs. But the sea is teeming with living plants.

Floating or drifting plants in the sea are part of the PLANKTON (PLANK-tun). Other members of the plankton are the protozoa and fungi. Of course, when we speak of food making, we are speaking of the green plankton, the algae. The ocean also contains the "big league" plankton. These form the large seaweeds which you have seen at the seashore. Some of them are "giants" and grow to be almost 300 feet! They may be red, brown or just green. The green ULVA (Fig. 128–1) looks like lettuce. There are giant kelps which are harvested to make a kind of fuel. There is the SARGASSUM which grows in the *Sargasso Sea.* (Find the Sargasso Sea in the Atlantic Ocean on a map. It is just northeast of Cuba.)

At the present time, the most important use of sea plants is the indirect one: it forms the food for fish which we use for our food. However, some countries, especially Japan, use seaweeds as vegetables. They are also "farming" an alga called CHLORELLA (klo-REL-uh) for food.

Fig. 128–1

Green sea plants carry on photosynthesis. It is estimated that over 80% of all the world's oxygen comes from plants in the ocean. How does oxygen reach you?

UNDERSTANDING WHAT YOU HAVE READ

I. What are the two main ideas of Chapter 20?

1. Green plants are good to eat.
2. Green plants make their own food.
3. Roots and stems are used for food.
4. The sun's energy is needed to make all the food in the world.

II. Explanation please . . . answer in one or two sentences.

1. What energy is needed for food-making in green plants?
2. Where do plants store extra food?
3. What are the raw materials needed in photosynthesis?
4. Where do animals get their food?

III. Knowing What and Why

1. Photosynthesis is possible
   a. only in light.
   b. only in soil.
   c. in light and dark.
   d. only in laboratories.
2. Bacteria and mushrooms cannot make their own food because their cells do not have
   a. calcium and iron.
   b. carbon dioxide.
   c. sun energy.
   d. chlorophyll.
3. Green plants get the raw materials for food-making from
   a. air and chloroplasts.
   b. soil and sun.
   c. air and soil.
   d. sun and water.
4. When we eat meat from a cow, we are eating food made by photosynthesis because
   a. cows carry on photosynthesis.
   b. cows eat green grass.
   c. meat is made from carbon dioxide and water.
   d. The statement is false.

IV. Match the food with its plant

1. carrots	a. seeds
2. celery	b. leaves
3. lettuce	c. fruits
4. beets	d. roots
5. melon	e. stems
6. plums	
7. grapes	
8. rice	
9. nuts	

V. Can you complete the story?

Human beings need food to live. They can use either _____ or _____ for food. Whichever they use, all food comes from _____ cells. If all green plants died, all _____ would soon die. Both plants and animals get their energy from _____. We make our meals different by using many parts of plants or animals. Chicken is a kind of _____. We can start our meal with a salad of shrimp, which is a _____.

a. green
b. plants
c. sunlight
d. poultry
e. shellfish
f. living things
g. animals

VI. For the Amateur Scientist

Make a collection of as many nuts as you can find. Use small jars for holding your samples. Print a label with a description of the plant from which each one is picked. Bring your collection to class.

TARGET
How do plants and animals get energy from food?

21

Chemical Change in Food Releases Energy

1. BURNING FUEL MAKES ENERGY

When the furnace of a steam engine is filled with coal, we know that there is no heat given off until the fire is lit and the coal is burned. Unless the coal is burned, there is no energy formed to make the steam that will drive the engine. In the automobile a full tank of gasoline is of no use until the engine burns it. Only then will the fuel (gasoline) produce energy to make the wheels move. In living things, the digested food is the fuel. Once again, this fuel is of no use to the living "engine" until it has been burned. Are you puzzled? Perhaps you are saying to yourself, "I've never seen a fire in my body." Of course not! But there are evidences that "burning" is taking place.

2. THE BURNING WE CANNOT SEE

Burning food in our bodies is not a fire such as you know, with light and

Fig. 130–1
Why is exhaled air warm? How can you measure its temperature?

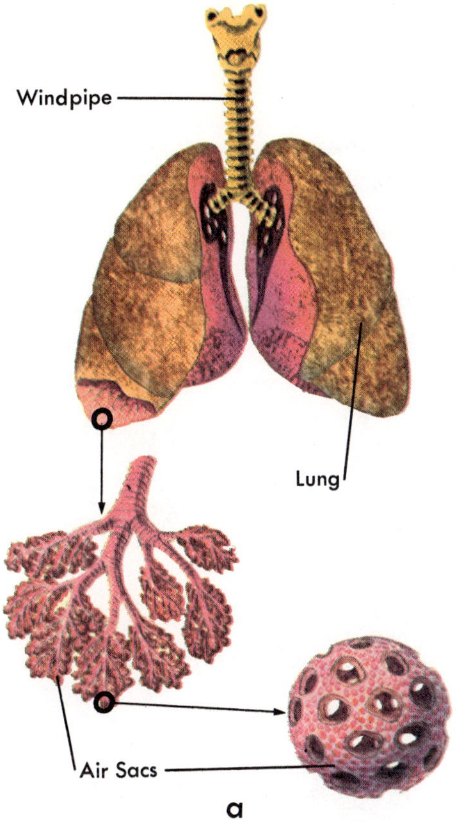

Fig. 131–1a

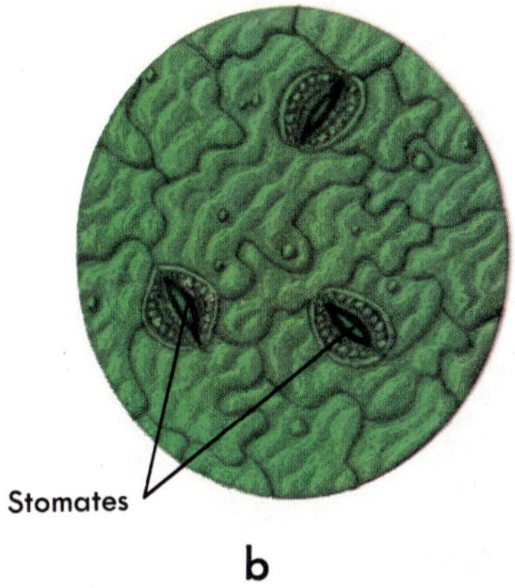

Fig. 131–1b
Living things have breathing organs.

flame. But we do know that heat is made by living things all the time. Here is a very simple experiment. Cup your hands and breathe into them. Is the air you breathed out warm or cold? You know from your experience in the wintertime that no matter how cold the air you breathe in is, the air you breathe out is warm. You have probably used this knowledge to warm your hands on a cold winter day. A quick way to check whether an animal is alive is to feel its body for warmth. Dead animals quickly turn cold.

3. PLANTS AND ANIMALS TAKE IN AIR

Perhaps you already know that burning requires the gas oxygen. You also know that our air is a mixture of gases. One fifth of the air is oxygen.

In Fig. 131–1a, you can see the organs used by one kind of animal to obtain the mixture of air with the oxygen it contains. And you know that we call this process *breathing*. Look at Fig. 131-1b. This pictures the underside of a plant's leaf. Here you can see the tiny openings, called *stomates*, through which the leaves take in air containing oxygen. Fig. 59–1a shows a breathing system in insects.

YOU NOW KNOW

- Living things need energy to do their work.
- Food is the fuel that gives energy as it is burned.
- To burn food, living things need to take in oxygen.

4. WHAT CHEMICAL ACTION MAKES ENERGY IN LIVING THINGS?

The small intestine is lined with thin blood tubes into which digested food is absorbed. In Fig. 131–1a, the lungs are also seen to be lined with thin blood tubes into which oxygen is absorbed. The blood then contains both digested food and oxygen.

Those who have studied chemistry know that burning is a chemical action in which something is combined with oxygen. The diagram below shows this process.

5. COMPARING OXIDATION AND BURNING OF FUEL

The chemical union of carbon or fuel with oxygen is called oxidation. Light and heat are given off and another material, carbon dioxide, is formed.

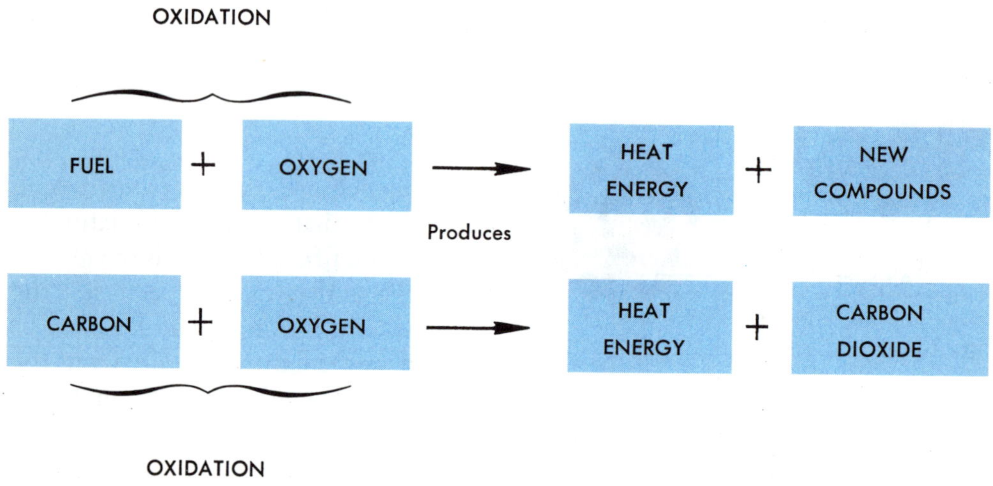

6. SLOW OXIDATION OF FOOD

Now let us draw a similar diagram for what goes on in a living thing:

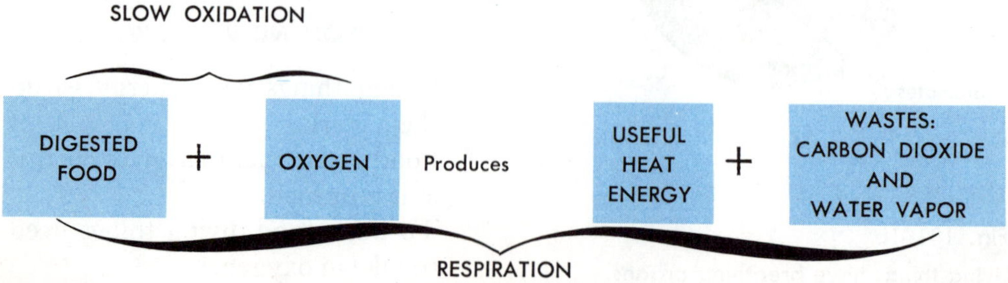

7. RESPIRATION—THE SOURCE OF ENERGY

This process, as you can see, is very similar to the combustion of fuel. There is one very important difference. The slow oxidation of food produces no light. The entire process is called RESPIRATION (res-per-AY-shun). Plants also absorb oxygen through their stomates. Then, through respiration, they oxidize their food to produce useful energy and heat.

Fig. 133–1

These Scouts walk hard and breathe deeply. How are they using the oxygen they breathe in?

YOU NOW KNOW

- Living things use oxygen to oxidize their food.
- This slow oxidation is called respiration.
- Respiration is a sign of life.

UNDERSTANDING WHAT YOU HAVE READ

I. What are the two main ideas of Chapter 21?

1. Food is a fuel which is oxidized to make energy.
2. Living things "burn" food to make energy.
3. Carbon dioxide is a product in breathing.
4. Stomates are like lungs.

II. In which paragraph do you find the answer?

1. What is breathing?
2. What kind of oxidation is used in respiration?
3. Where does oxygen enter our blood?
4. What wastes are made during respiration?

III. Knowing What and Why

1. Oxygen is needed by living things
   a. to create carbon dioxide.
   b. to allow for breathing.
   c. to oxidize food.
   d. to create wastes.
2. Plants absorb oxygen
   a. through the stomates in their leaves.
   b. through their roots.
   c. through their stems and bark.
   d. even though it is not necessary.
3. Respiration is a chemical process which
   a. has nothing to do with breathing.
   b. has nothing to do with oxidation.
   c. includes breathing and oxidation.
   d. is the opposite of excretion.
4. Excretion from the lungs
   a. avoids poisoning the body with wastes.
   b. makes room for more air.
   c. helps circulation.
   d. helps sweating.

IV. Match the terms correctly.

1. waste gas	a. excretion
2. removal of waste	b. leaf openings
3. oxidation	c. lungs
4. breathing	d. chemical change
5. stomates	e. carbon dioxide
	f. bromothymol blue

V. Which comes first? Arrange these in order.

1. digesting—eating—circulating—excreting
2. oxidation—respiration—breathing—sweating

VI. Using Your Judgment—Can you explain?

1. Why do people breathe faster on a mountaintop than in a valley?
2. Why do people breathe on their glasses before they polish them?

VII. Put on Your Thinking Cap . . .

High up in the Andes Mountains in Peru, the native Indians have about seven quarts of blood to our five quarts. Their chest cavities are generally larger than ours, so are their lungs.

1. Why would you find such a condition in high mountains?
2. What idea does this give you about the need for oxygen in high mountains?
3. What idea does this give you about the use of extra blood?

22

TARGET
Where do plants and animals get their food?

The Search for Food and Nutrients

Fig. 135–1

1. WHAT DO WE EAT?

Let us think of some common foods. We eat steak, chicken, fish, eggs, potatoes, apples, oysters, onions, lettuce, cheese, pork, bread and corn. All animals eat plants or other animals. There are some animals, like elephants and chickens and horses, that eat only plant materials. Other animals, like the lion or fox, eat only meat (flesh). Human beings eat both plants and animals. In Chapter 20, paragraph 6, we saw which parts of green plants are used for food. Look back at the chart on page 127.

2. WHAT IS A FOOD?

In Unit II, Chapter 8, pages 51–52, we saw that living things take in food. We saw that when food is digested, it can then be "burned" to give us the energy to continue to live. Later, we used the words *growth* and *assimilation* to show how food is made into living protoplasm. Let us now review: food is what we eat to give us energy, to make us grow, to repair cuts and injuries, and to protect us from certain diseases.

Food is for → Energy, Growth, Repair, Protection

3. ANIMALS FOR FOOD

Many animals, including man, use other animals and animal products for food.

Animals Provide → Meat, Eggs, Dairy Products {Milk, Cheese, Butter, Cream}

The most common animals used for their meat are cattle (beef), sheep (for lamb and mutton) and hogs (for pork, ham and bacon). But other animals are also used as food. These include deer, rabbits, goats, frogs, snakes and lizards. Chicken, turkey and duck are the most common birds eaten for their flesh. We call their group POULTRY (POLE-tree).

4. FOOD FROM THE RIVERS AND OCEANS

The rivers and oceans are filled with fish, many of which have excellent, tasty flesh. Some of them, such as the cod, are used to make oil. Others, like the shad, have tiny eggs which we call ROE (ROW). The

Fig. 136–1
Typical seafood.

eggs of the sturgeon are the very expensive CAVIAR (KAV-ih-yar).

Shellfish are animals such as lobsters, shrimps, crabs, clams and oysters. SEAFOOD is any animal food taken from any water—not only from the sea.

YOU NOW KNOW

- All food comes from green plants.
- Animals eat plants or other animals.
- Man eats many parts of plants and many parts of animals.

5. WHAT ANIMAL PARTS ARE USED FOR FOOD?

Usually we eat the muscle parts of the animal. Their flesh may also have skin and fat attached to it. The words *steak*, *chops* and *spare ribs* only tell us which way the butcher cut the meat. Some inside organs of animals are also used for food. These are known as ORGAN MEATS. Examples of organ meats are lungs, liver, tongue and kidneys.

Fig. 137–1
This scientist is testing well water for pollution.

6. CLEAN WATER IS A NEED OF MAN

Sixty-seven percent of living matter is water. To keep cells working correctly, the amount of water must be kept about the same. There are cases where people have been without food for forty days and have continued to live, but this is very unusual. Failure to have water for more than ten days generally means death.

However, it is not enough to have water. Water must be clean, not POLLUTED (puh-LOO-ted). Polluted water is dirty water. It gets dirty as dead plants and animals are washed into it. Or it may get dirty because factories carelessly dump wastes into it. Polluted river water kills the fish which live in it. It is also unfit to drink because it smells bad, tastes bad and may carry disease germs. Modern man must find ways to keep water clean. Look at the diagram in Fig. 138–1. It shows a laboratory method to remove undissolved waste from water. We call this process FILTRATION (fil-TRAY-shun).

Government and industry are now looking for ways to stop dumping wastes into rivers. In this way, we can prevent water pollution. Water is also treated with other chemicals to destroy odors and improve its taste.

CAUTION

Never go swimming in polluted water!
Never drink water unless you are sure it is pure.

DO AND DISCOVER

Investigation 16: How can filtering make water pure?

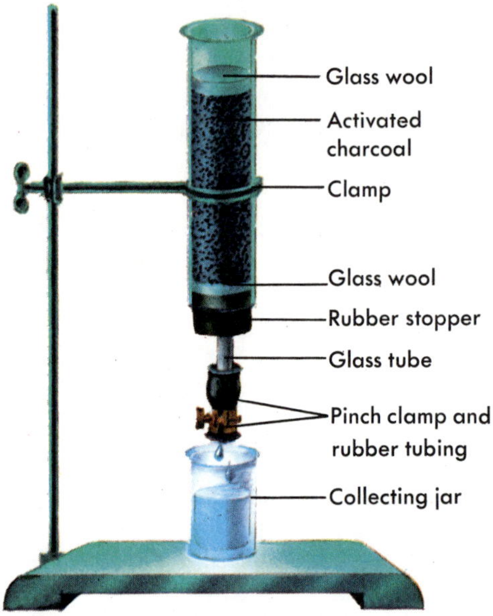

Fig. 138–1

Can you explain why the blue water will appear colorless when it drips into the collecting jar?

Fig. 138–2

Chlorine gas is added to the water supply to kill disease germs.

7. THE "PAYLOAD" IN FOOD

If you have been reading the newspaper stories about space rockets, you may have come across the word "payload." A rocket is made of several parts. Some parts are used only to get the capsule into space. These parts drop away early in the flight. The part which is left—the capsule that carries the spacemen—is called the "payload." The foods we eat have many useless compounds which are given off in excretion. But some parts of the food are not lost in excretion. Some stay in the system to complete the digestive process, just as the "payload" capsule completes the rocket flight. These parts of the food are finally used by the body to carry on life functions. These useful parts are called NUTRIENTS (NEW-tree-ents). Nutrients are the "payload" in food.

8. THE MEANING OF NUTRIENT

A nutrient is the part of a food which can be used for energy, for growth or for repair of damaged tissues. What are nutrients? They are chemicals—mostly compounds, but some elements as well.

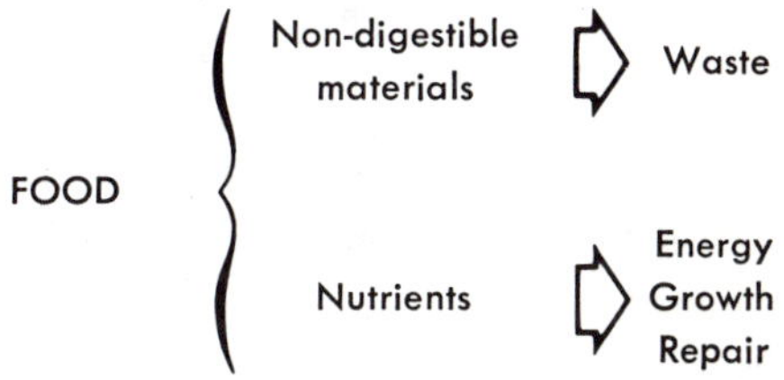

Most nutrients have been tested in the laboratory, and we know what they are made of. Most nutrients can be made only by green plants. A very small number can be made only by animals. In the next chapter, we shall learn more about nutrients.

UNDERSTANDING WHAT YOU HAVE READ

I. What are the two main ideas of Chapter 22?

1. Animals get their food from other animals or plants.
2. Fish is just as good a food as meat.
3. Polluted water makes you sick.
4. We can use only the nutrient parts of a food.

II. In which paragraph do you find the answer?

1. What are the most important uses of food?
2. What group of animals is called poultry?
3. What are organ meats?
4. What do we call the useful part of a food?

III. Knowing What and Why

1. Which of the following groups are organ meats?
   a. skin, steak and liver
   b. kidneys, liver and tongue
   c. fish, lungs and lobster
   d. crabs, chops and steak
2. Filtration of polluted water aims to
   a. remove germs.
   b. remove waste.
   c. add oxygen.
   d. add fluorine.
3. Assimilation is a process of making protoplasm from
   a. water.
   b. nutrients.
   c. oxygen.
   d. growth.
4. The foods we eat weigh more than the nutrients we use because
   a. wastes are removed in digestion.
   b. assimilation changes the weight.
   c. oxidation changes the weight.
   d. foods contain water.

IV. Match the words in Column A with the words in Column B

A	B
poultry	fish eggs
seafood	duck
caviar	cream cheese
organ meat	oyster
dairy product	lamb chop
	whale
	kidneys

V. For the Amateur Scientist

1. Visit a butcher. Ask him to point out different "cuts" of meat.
2. Visit a fish market. See the different types of food for sale. How many can you recognize?

VI. Unscramble the letters. Then list the first letter of each new word. You will get the first name of a famous boxer.

1. RBYRADACOHET (a food nutrient)
2. ATLINOIMIASS (changing food to bone)
3. OISMUD (an element in table salt)
4. AEFSODO (fish, clams, and so forth)
5. GESTININO (taking in food)
6. RAUE (a waste product)
7. UGRAS (compound used as food)

Now write the first letter of each word. What name do you get?

23

TARGET
How can we tell which nutrients are found in a food?

Identifying the Nutrients

1. THE FOOD DETECTIVES

Did you ever hear of a food detective? Of course not. But a chemist can be a food detective in a way. By using his skills, he can detect (search out) what is really in a food.

2. SIX GROUPS OF NUTRIENTS

Chemists have found six groups of nutrients in foods:

1.	**CARBOHYDRATES**	Sugars; Starch	Compounds of carbon, hydrogen and oxygen
2.	**FATS**	Oils; Solid Fats	Compounds of carbon, hydrogen and oxygen
3.	**PROTEINS**		Compounds of carbon, hydrogen, oxygen, sulfur, phosphorus and nitrogen
4.	**MINERALS**		Examples: iron, calcium, sodium and iodine
5.	**VITAMINS**		Compounds of carbon, hydrogen oxygen, and nitrogen
6.	**WATER**		Compound of hydrogen and oxygen (H_2O)

3. TESTS FOR DISCOVERING NUTRIENTS

How can we tell whether potatoes have starch in them? Or whether milk has protein or sugar or both? Scientists have invented certain NUTRIENT TESTS. These are chemical tests which are very easy to perform. You can do many of them yourself at home.

To understand how these tests work, let us imagine that we have the nutrients in *pure form.* In other words, we have removed the nutrient from the food. In each case, we shall do the test twice: *first*, on the pure nutrient; *second*, on a food that contains the nutrient.

YOU NOW KNOW

- Food consists of wastes and useful chemicals called nutrients.
- Nutrients are chemical compounds belonging to one of six groups.

4. IS IT STARCH?

DO AND DISCOVER

Investigation 17: How can we test for the presence of starch?

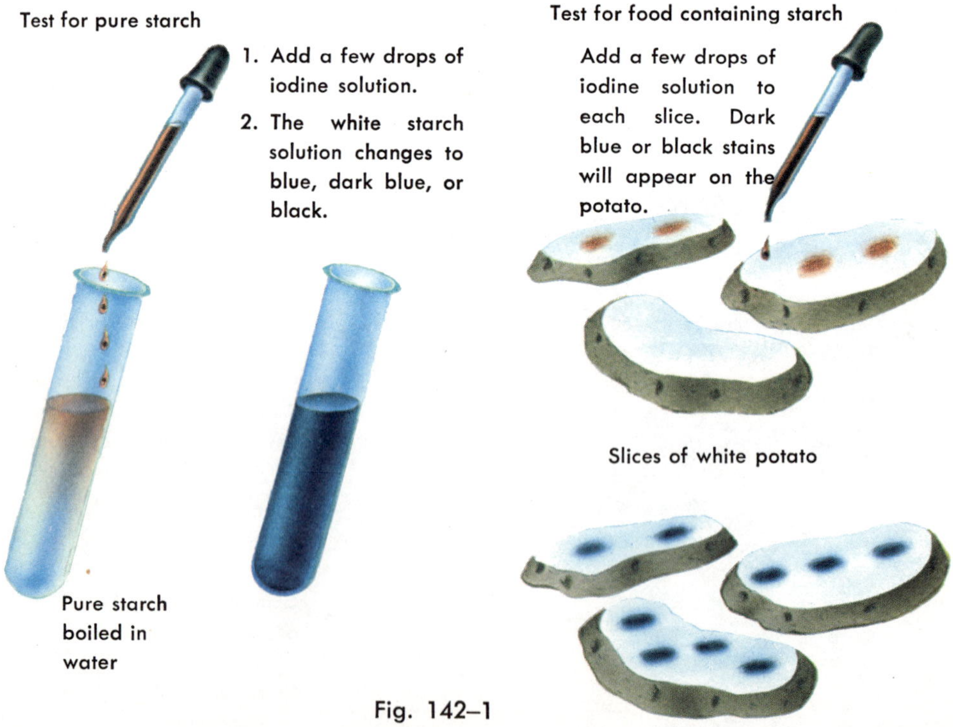

Fig. 142–1
The test for starch. Would this test work on rye bread? Explain.

YOU NOW KNOW

- Adding iodine to starch makes a dark blue color.
- A potato will turn dark blue or black when iodine is added.
- Therefore, we know that the potato is one food which contains starch.

5. IS IT SUGAR?

DO AND DISCOVER

Investigation 18: How can we test for the presence of sugar?

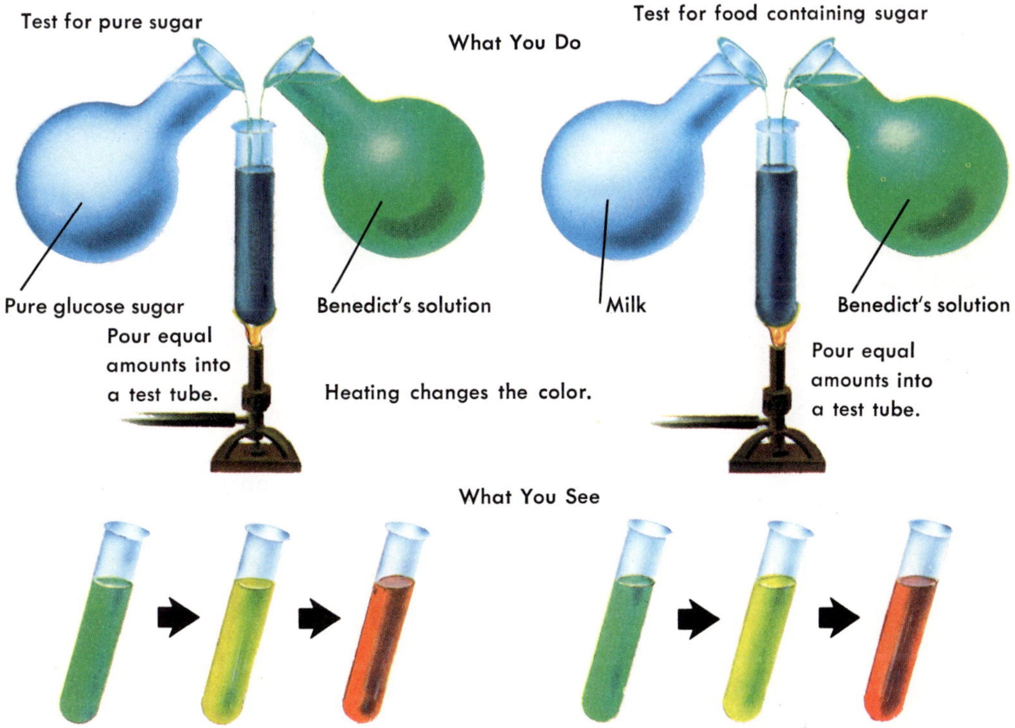

Fig. 143–1
The test for sugar. Explain each step.

YOU NOW KNOW

- Sugar mixed with Benedict's solution and heated changes from green to yellow and then to red.
- This test will tell whether a food contains sugar.

6. IS IT PROTEIN?

There is a pure protein called CASEIN (KAY-seen) which we can use for this test. The test for protein can be made easily, as shown below. The chemical solution we use is called BIURET (by-a-RET) solution.

DO AND DISCOVER

Investigation 19: How can we test for the presence of protein?

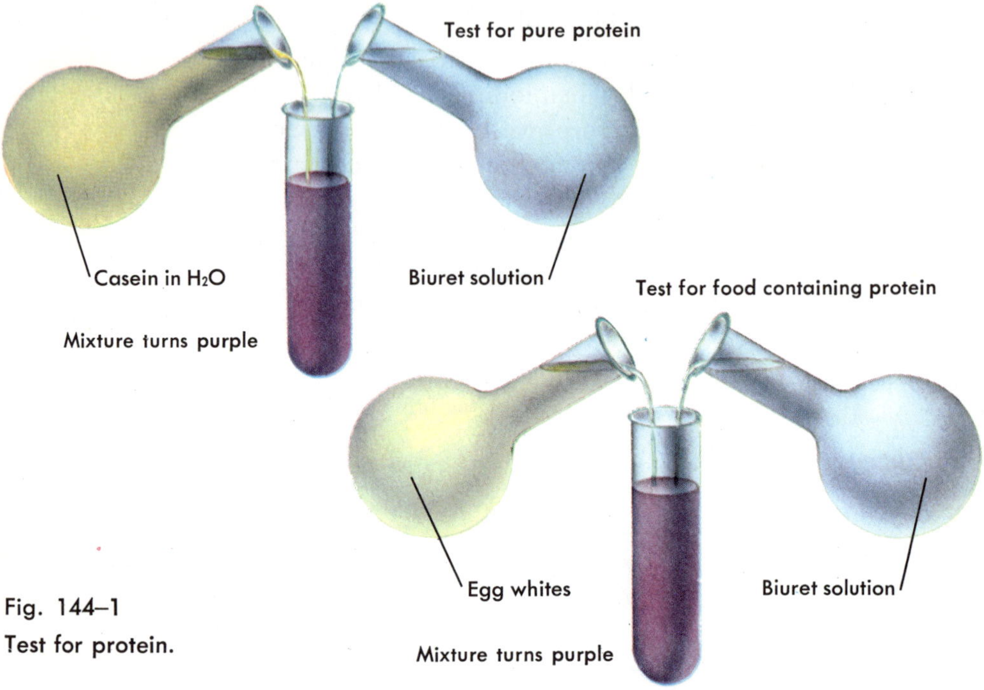

Fig. 144–1
Test for protein.

YOU NOW KNOW

- When Biuret solution is added to protein, a purple color results.
- When Biuret solution is added to egg white, the mixture turns purple. Therefore, we know that egg white contains protein.

7. IS IT FAT?

Moisten your finger with a drop of oil. Rub your finger across a piece of brown wrapping paper. Leave it to dry. Hold the paper up to the light. There will be a "grease" spot on the paper that lets the light shine through. Now rub another piece of brown paper with a small pat of butter. Leave it to dry. Hold the paper up to the light. There will be a "grease" spot on this paper similar to the first. This is a simple test for fat.

8. IS IT A MINERAL?

When a small piece of white bread is heated in a clay dish, the material which is left after the bread is completely burnt is a white or gray ash. The same thing happens when you heat a peanut. This is a simple test for minerals in food. Calcium, iron and iodine compounds are examples of minerals found in foods. Of course, other tests are needed to decide which mineral has been found.

DO AND DISCOVER

Investigation 20: How can we test for the presence of minerals?

Fig. 145–1

Test for minerals.

9. IS IT VITAMIN C?

There are several vitamins, popularly called by letters: vitamins A, B, C, D and K. At least one of them, vitamin C, can be tested easily.

As you add the drops of pure vitamin C, the blue INDOPHENOL (IN-do-FEE-nul) begins to turn clear (colorless). As you add the drops of lemon juice, the blue indophenol begins to turn clear (colorless).

YOU NOW KNOW

- Blue indophenol turns colorless in the presence of vitamin C.
- Lemon juice is one food which contains vitamin C.

10. DOES FOOD CONTAIN WATER?

All foods have water. Some have only a little; some have a great deal.

DO AND DISCOVER

Investigation 21: How can we test for the presence of vitamin C?

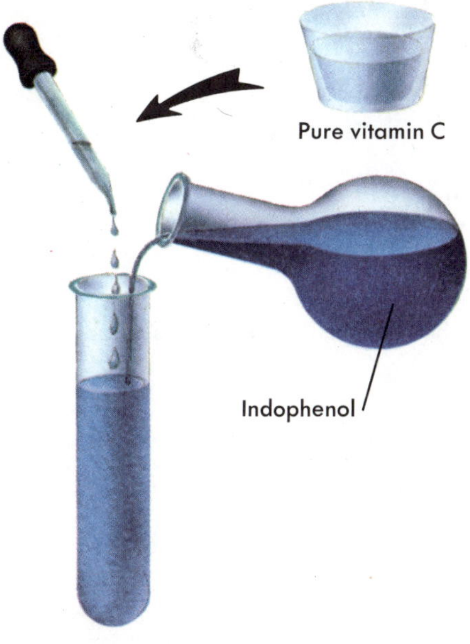

Fig. 145–2

Test for pure vitamin C.

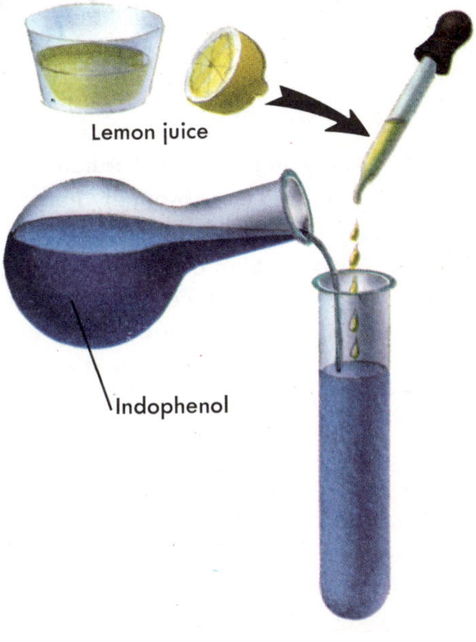

Fig. 145–3

Test for food containing vitamin C.

There is an easy way to prove this. When you heat a food in a dish, a vapor leaves the food. If you hold a cool piece of glass above the heated food, the vapor will change to drops of water on the glass.

YOU NOW KNOW

- There are some simple tests to identify pure nutrients.
- These same tests tell us which nutrients are found in a food.

DO AND DISCOVER

Investigation 22: How can we test for the presence of water?

Fig. 146–1
Test for water.

Analyzing the Investigation

1. Is the heat absolutely necessary? Explain.
2. If you don't have a gas burner, what else might you use?
3. What kind of dish should you be using? Why?
4. Why might we first weigh the food we are using?
5. In this investigation only plant food is shown. Could you use a similar investigation for meat or fish?
6. Why are tongs needed?

UNDERSTANDING WHAT YOU HAVE READ

I. What are the two main ideas of Chapter 23?

1. Starch, fats and protein are life-giving chemicals.
2. Nutrient tests tell us which parts of the foods we eat are valuable.
3. There are chemical tests to identify nutrients.
4. Some foods have more nutrients than others.

II. Explanation please . . . answer in one or two sentences.

1. What minerals are found in food?
2. For which two nutrient tests do we *not* need chemicals?
3. Is there water in food?
4. What chemical elements are found in sugar?

III. Knowing What and Why

1. Two kinds of carbohydrates are
   a. protein and sugar.
   b. fats and minerals.
   c. starch and sugar.
   d. vitamins and starch.
2. The nutrient minerals are usually
   a. traces of chemical elements.
   b. compounds with nitrogen.
   c. water plus sulfur.
   d. the same as vitamins.
3. If milk is tested for sugar, fat and protein, and the answer is "yes" to all three, then
   a. the milk is spoiled.
   b. the tests were done wrong.
   c. milk does have these three nutrients.
   d. it is a special kind of milk.
4. The nutrient tests depend on color changes, except
   a. sugar and minerals.
   b. minerals and water.
   c. protein and starch.
   d. sugar and starch.

IV. Match the words with the colors.

1. Biuret solution and protein	a. blue-black
2. indophenol and lemon juice	b. orange-red
3. starch and iodine	c. purple
4. Benedict's solution and sugar	d. blue to colorless
	e. green

V. Can you complete the story? Use the words below.

All foods contain one or more _____ which are useful for energy, growth, and repair. To find out which nutrient a food has, we use _____ which usually show a _____. There are _____ groups of nutrients, which are placed in different groups because they contain different __________. Sugar and starch are very much alike and are sometimes put into one group called __________.

chemical change
tests
six
carbohydrates
nutrients
color change
eight
compounds

VI. For the Amateur Scientist

1. You can do the starch-iodine test very easily at home. Add a few drops of iodine to white bread, rice, potato and cracker.
2. Buy a package of *Clinitest* pills in the drugstore. Drop one pill into a small jar of syrup. These pills will "fizz" and cause the syrup to show the presence of sugar (green and orange-red).

24

TARGET
Which foods shall we eat for better living?

Eat Right . . . Live Right

1. WHAT'S IN THIS FOOD?

You have already learned that by testing foods we can discover what nutrients they have. But that only tells part of the story. We know a potato contains starch. But how much starch? We know milk contains sugar. But how much sugar? And what other nutrients are in milk? Scientists have tested foods and know these answers. They have arranged their results in *food tables* or *food charts*.

2. WHAT DOES A FOOD TABLE SHOW US?

Look at Fig. 149–2. This is a list of common foods showing the percentage (how many parts out of 100) of carbohydrates, fat and protein each food contains. Such tables are useful to people who plan diets. Perhaps some of you, in the future, would like to be DIETICIANS (dy-a-TISH-unz) and plan diets for schools, hospitals or factory lunchrooms. In that case, a food chart would come in very handy. For example, you can see that apples have very little protein (less than 1%), while fish and eggs are very rich in protein. Examine the line for milk. Can you now see why milk is sometimes called a "perfect food"? Fig. 150–1 is a table of common foods and their water content. Fig. 149–1 shows mineral contents and Fig. 152–1 shows vitamin contents.

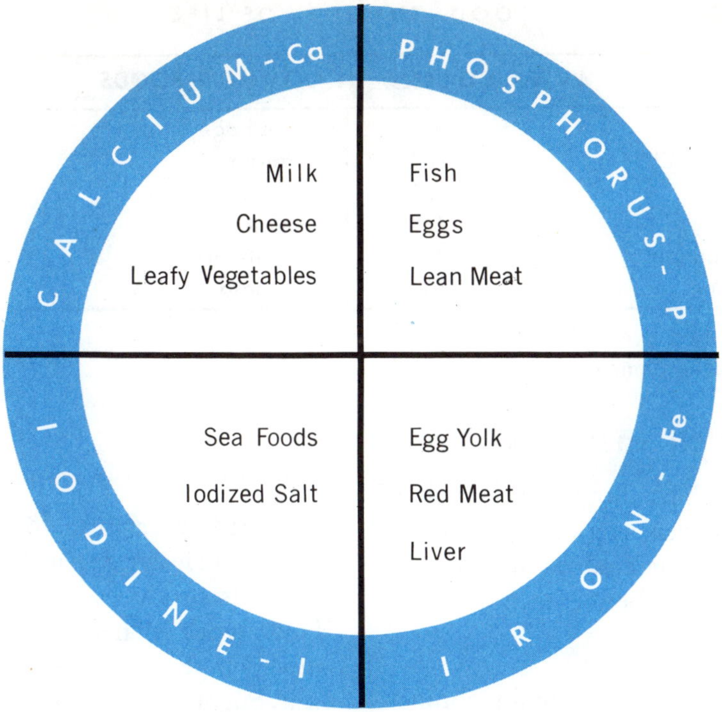

Fig. 149–1
Foods for minerals.

NUTRIENTS IN COMMON FOODS

Average Serving	Protein	Fat	Carbohydrate
Apple	*	*	14%
Bacon	10%	64%	—
Beans	5%	*	22%
Beef (hamburger)	19%	12%	—
Bread (white)	10%	4%	57%
Butter	*	80%	—
Cabbage	1%	*	6%
Cheese, cottage	20%	*	2%
Eggs	12%	9%	—
Fish	21%	12%	—
Ham	16%	37%	—
Ice Cream	4%	13%	20%
Lettuce	*	*	3%
Milk	4%	4%	5%
Peas	5%	7%	17%
Potatoes	2%	*	20%
Rice	2%	*	24%

Fig. 149–2

* Less than 1%. Reminder: one percent = 1 out of 100 parts.

WATER CONTENT IN COMMON FOODS			
Milk	87%	Cabbage	92%
Cottage Cheese	76%	Potato	74%
Bacon	13%	Apple	84%
Raw Egg	74%	Watermelon	92%
Hamburger	47%	White Bread	35%
Salmon	70%	Tomato	94%

Fig. 150–1

Water in Common Foods.

YOU NOW KNOW

- Food tables show how much of each nutrient is found in foods.
- By using these tables, we can plan our diet for better health.

3. HOW ARE NUTRIENTS USED?

If you were a mountain climber, you would want to take a very small package of food that would give you a great deal of energy to climb long and hard. If you wanted to gain weight, increase your muscles and make your bones strong, you would look for foods that could build protoplasm. It is easy to understand then why the six nutrient groups are sometimes placed into two larger groups.

4. LOOKING BACK

Reread paragraph 6, page 132. There you learned that slow oxidation of foods produces heat and energy. To do work we need energy. To keep our body temperature even, we need heat. Sugars and fats burn rapidly and give off heat and energy. It is not surprising that candy is often carried by hikers to give quick energy. People like the Eskimos, who live in the frozen northern countries, eat great quantities of fats. Fats and other nutrients which are eaten but not used are stored in the body as fat. The liver and the muscles can store sugar for future use.

1. ENERGY NUTRIENTS
   1. carbohydrates 2. fats
2. BUILDING NUTRIENTS
   1. water 2. proteins 3. minerals 4. vitamins

Starchy Foods	Sugar Foods	Fatty Foods
Potato	Sugar	Butter
Bread	Candy	Margarine
Rice	Chocolate	Bacon
Cereals	Cake	Cream
Spaghetti	Banana	Olive Oil
Noodles	Ice Cream	

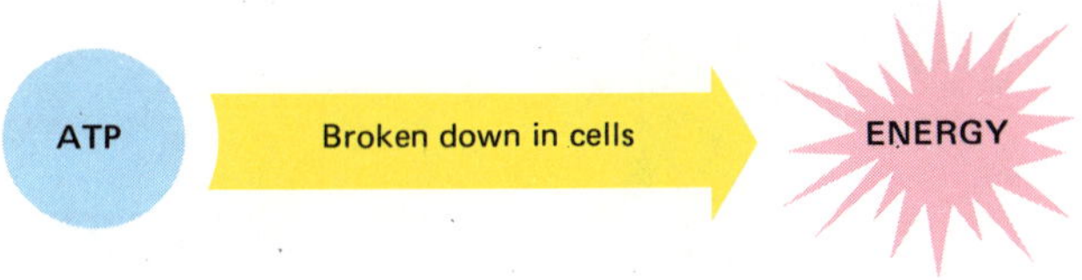

5. ENERGY FROM WHAT? ENERGY FROM WHERE?

Burning coal can give energy to an engine; so can gas; so can gasoline. What is the material in living things that provides energy? It comes from the oxidation of food. But how? The answer is an amazing compound. It is a protein-like compound which also contains compounds of *phosphorus* (FOS-fur-us). We call it ATP for short.

The cells build ATP from food. The ATP stores energy. Then the cells break down ATP to give off energy and heat. And then? Of course! Your body needs more ATP. Again, it uses food to build ATP. This is another example of a *cycle* in living things.

6. NUTRIENTS FOR BUILDING

When you studied protoplasm and tissues, you learned that all living matter is made from chemicals in our food. Proteins are changed by cells into muscle, flesh and skin. The mineral iron, as you saw, is needed for building red blood cells. The minerals calcium and phosphorus are needed for strong, healthy bones. The mineral iodine is needed to keep the thyroid gland working. When we think of protein foods, we think of meat, fish, eggs and milk. Water is the basic fluid of all protoplasm. Since you lose water through perspiration and excretion, you should drink a quart of water every day. See Fig. 150–1, which tells you how much water you get from common foods.

YOU NOW KNOW

- Proteins, minerals and water are nutrients used to build cells, protoplasm and body parts.

Fig. 152–1
Why do all these people need large amounts of energy foods? Where in their bodies is this energy needed?

7. VITAMINS FOR BETTER HEALTH

Did you know that a person can eat enough to satisfy his appetite, and yet be "starved"? About three hundred years ago, sailors who came back from long sea trips often had a disease called SCURVY (SKUR-vee). Their gums would bleed. They would feel weak and run-down. Someone discovered that eating limes (a fruit like the lemon) or drinking lime juice would cure scurvy. Later, scientists found out that foods contain very small amounts of chemicals called VITAMINS. They are named by letters (A, B, C and so forth). We need only the tiniest amounts of vitamins. But if we don't have enough vitamins, we can actually "starve" or get sick. Look at the chart in Fig. 153–1. You will see which foods contain vitamins and what might happen to your body if a certain vitamin were missing from your diet.

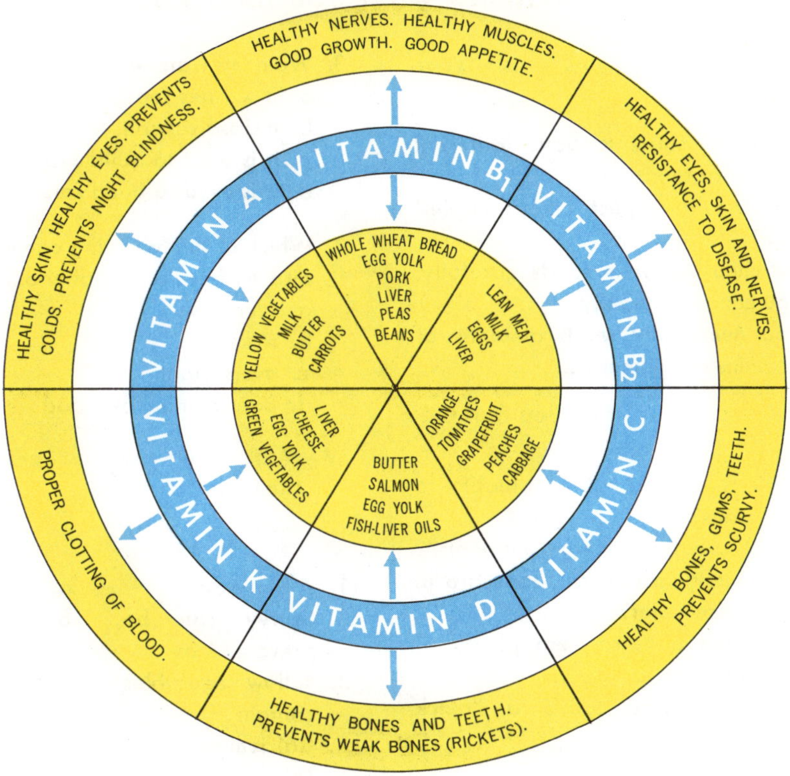

Fig. 153–1

"The Vitamin Wheel"

The inside wheel shows which food supplies the vitamin.

The outside wheel shows its use to the body.

YOU NOW KNOW

- Vitamins are chemicals which keep our bodies working correctly.
- You need only small amounts of vitamins.
- Most foods have some vitamins in them.

UNDERSTANDING WHAT YOU HAVE READ

I. What are the two main ideas of Chapter 24?

1. Different nutrients have different values for the body.
2. We should respect other peoples' foods.
3. By choosing your foods carefully, you can stay healthy.
4. You need only small amounts of vitamins and minerals.

II. In which paragraph do you find the answer?

1. Why is a food table valuable to dieticians?
2. What are the "energy nutrients"?
3. How many groups of foods are in a balanced diet?
4. Why can a person eat plenty of food and yet "starve"?
5. Why is candy called a "quick energy food"?

III. Knowing What and Why

1. Seafoods can supply
   a. the mineral iodine.
   b. vitamin A.
   c. sugar.
   d. vitamin B.

2. Of the following, which group is richest in protein?
   a. fish, bread, potato
   b. fish, eggs, jelly
   c. butter, milk, spaghetti
   d. fish, eggs, steak

3. We need water in our diets because
   a. it washes food down.
   b. it dissolves salt.
   c. it is used to build protoplasm.
   d. it is used to build fats.

4. Which of the following cannot result directly from a lack of vitamins?
   a. scurvy
   b. rickets
   c. overweight
   d. poor clotting of blood

IV. Match the words in Column A with those in Column B.

A	B
1. A dietician	a. vitamin C.
2. Scurvy can be prevented by	b. oxidation.
3. Yellow vegetables contain	c. vitamin A.
4. Calcium builds	d. plans diets.
5. The release of energy is	e. milk.
6. The "perfect food" is	f. vitamin D.
7. The "sunshine" vitamin is	g. good teeth.
	h. assimilation.
	i. Biuret.

V. For the Amateur Scientist

1. Collect labels from food packages. Paste them on a chart. Make a summary of what each one contains.
2. Make a list of all your meals for three days. Using the nutrient chart, figure out the amount of each nutrient you had each day.

TARGET
How do we count Calories?

25

Counting Calories

1. HOW MUCH ENERGY?

Which do you think has more power on the road—a small foreign car or a big American car? We measure power of cars in "horsepower." A small foreign car may have only 40 horsepower. A big American car may have 350 horsepower. When we talk of energy needs for human beings, we do not speak of "horsepower"; we speak of "Calorie energy."

Fig. 155–1
Who needs more "Calorie energy"?

2. MEASURING THE ENERGY IN FOOD

In the previous chapter, you learned that some foods are richer in one nutrient than in others. The amounts and kinds of nutrients in a food determine how much energy the food will give you. It is not enough to weigh the *pounds* of food you eat. A pound of meat would give you less energy than a pound of sugar. To understand how to measure the energy of foods, look at the experiment in Fig. 156–1. Scientists have found ways to measure the amount of heat energy given off when foods are "burned" in your body. This energy is measured in CALORIES (KAL-o-reez). You need energy to run, to work, to make your heart beat, and to breathe. Your body can also take in extra Calories which are not immediately needed for work, and store them as fat for later use. If you do not eat enough, your body begins to use up stored fat, and you then lose weight.

DO AND DISCOVER

Investigation 23: How can we measure the energy in foods?

Fig. 156–1

How can you figure out the number of Calories given off by the burning peanuts? What are some defects in this experiment?

3. CALORIES FOR ENERGY

How many Calories does a food have? This varies, of course, according to the food. It also depends on the size of the portion. Here are some examples:

CALORIES IN AVERAGE PORTIONS

Food	Calories
Corn Flakes (cup)	80
White Bread (slice)	75
Chocolate Bar (small)	350
Jello (cup)	75
Ice Cream (½ cup)	160
Butter (1 pat)	100
Egg (one)	70
Milk (glass)	170
Apple	100
Orange	80
Hamburger	100
Lamb Chop	211
Chicken (2 slices)	120
Baked Beans (½ cup)	115
Lettuce	5
Tomato	35

4. HOW MANY CALORIES?

Perhaps you are now asking, "How many Calories does a person need?" Look at the charts on page 157.

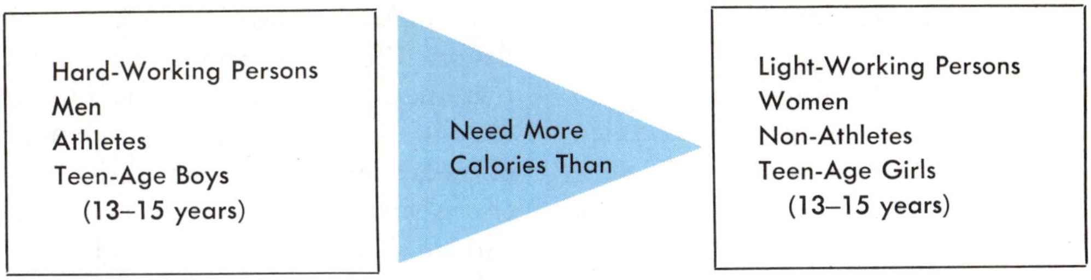

CALORIE NEEDS PER DAY

Hard-Working Man	4500
Non-Active Man	3000
Hard-Working Woman	3000
Non-Active Woman	2500
Teen-Age Boy	3200
Teen-Age Girl	2800
Children 1–6	1200–1600

Fig. 157–1
These figures show how many Calories are needed per day by different people.

A hard-working laborer who uses a pick and shovel all day will need more energy than a man with a desk job. A basketball player needs more energy than someone who is not an athlete. Teen-agers are very active and are growing very fast. They need much energy. Of course, large people need more food than small people do.

Fig. 157–2
Which workers need more Calories? Can you explain why?

5. IS A BEVERAGE A FOOD?

Most people enjoy something to drink with their meal. This is called a BEVERAGE (BEV-er-ij). The most common beverages are milk, cocoa, coffee, tea and soda. Milk, as you know, is a food in itself. When cocoa or chocolate is added, the milk becomes even more valuable as a food. Coffee and tea have no food value, however, except for the cream or sugar that is added. Both coffee and tea contain mild drugs which "pep" you up. Most sodas contain sugar and flavor. Teen-agers should be careful that they do not drink soda *instead* of milk. Also, the large sugar content of soda may harm your teeth. Beverages such as wine, liquor and beer contain ALCOHOL (AL-ko-hol). Alcohol is a dangerous and habit-forming drug. It affects the nerves and damages the kidneys and the liver. People who are drunk may become involved in accidents and crime. Heavy drinking may also lead to unemployment.

Fig. 158–1

Which beverage is most healthful? Explain. Do all of these contain nutrients? Explain.

YOU NOW KNOW

- Calories are units of heat energy in foods.
- The Calorie value of most common foods is known.
- You can figure your own Calorie needs for each day.

UNDERSTANDING WHAT YOU HAVE READ

I. What are the two main ideas of Chapter 25?

1. The number of Calories needed in a person's diet depends on age, weight, and activity.
2. All Calorie needs are approximate.
3. The energy in food is measured in Calories.
4. Extra Calories are useless.

II. Explanation please . . . answer in one or two sentences.

1. What are the Calorie needs of teen-agers?
2. What food value is there in tea and soda?
3. What happens to extra Calories not used for energy?
4. Compare the Calorie content of sweet, starch, fat, and protein foods.

III. Knowing What and Why

1. In general, high-Calorie foods are those with large amounts of
   a. proteins.
   b. fats.
   c. vitamins.
   d. calcium.
2. Teen-agers need about
   a. 3000 Calories per day.
   b. 4000 Calories per day.
   c. 2000 Calories per day.
   d. 5000 Calories per day.
3. To reduce your weight, you would need
   a. a high-protein diet with plenty of fats.
   b. a liquid diet.
   c. milk, water, minerals and vitamins only.
   d. a low-Calorie diet and exercise.
4. When we eat extra Calories, they are not used for energy. Instead our bodies may
   a. change them to vitamins.
   b. store them as fat.
   c. excrete them.
   d. get sick.

IV. Count the Calories

1. *Breakfast:* 2 eggs, fried in 1 pat of butter
   1 slice of white bread
   1 pat of butter
   1 glass of milk
2. *Lunch:* 2 hamburgers, each in a roll
   1 portion of ice cream
   1 glass of milk

V. On the Ladder of Understanding

Complete the story. Use the words below.

Calories measure __________ energy. Before the energy can be used, the useful __________ must be removed from the food by __________. The food is then __________ with oxygen in the process of __________. Ounce for ounce, foods that contain __________ also contain the largest number of Calories. Although coffee and tea are not really foods, the __________ which is added may make them useful.

digestion
respiration
fats
heat
oxidized
starch
nutrients
sugar
growth
excretion

TARGET
Shall I diet?

26

Controlling Your Diet

1. WHAT IS A BALANCED DIET?

Dieticians speak of a BALANCED DIET. Such a diet has all the nutrients necessary for good health. It is not enough to stuff yourself with food. You must eat the right kind of food and in the right amounts. The chart on the next page shows the *Basic Four Groups* from which foods should be picked each day.

2. IS YOUR WEIGHT RIGHT?

Some people are tall; some are short. Some people are heavy; some are light. Some people have big frames, while some have small frames. Whether your weight is "right" is very hard to tell. But there is an average to which you can compare yourself.

AVERAGE WEIGHT IN POUNDS BY AGE

Age	Boys	Girls
11 years	87	83
12 "	92	88
13 "	100	95
14 "	103	100
15 "	120	110
16 "	138	115
17 "	145	125
18 "	150	127

What other factors are not shown in this table that could influence weight?

3. SHOULD YOU GAIN WEIGHT? SHOULD YOU LOSE WEIGHT?

In general, if you feel well, if you have "pep" and if you look well, all you need to do is be sure you have a balanced diet. If there is a question, you should have a doctor advise you. As you learned before, diets heavy

Use Daily:

Milk Group

3 or more glasses milk — Children
smaller glasses for some children under 8

4 or more glasses — Teenagers

2 or more glasses — Adults

Cheese, ice cream and other milk-made foods can supply part of the milk

Meat Group

2 or more servings

Meats, fish, poultry, eggs, or cheese—with dry beans, peas, nuts as alternates

Vegetables and Fruits

4 or more servings

Include dark green or yellow vegetables; citrus fruit or tomatoes

Breads and Cereals

4 or more servings

Enriched or whole grain
Added milk improves nutritional values

Fig. 161–1
For health and energy, pick foods from all four groups.

in starch, sugar and fat will cause a person to gain weight. People who wish to reduce will therefore eat less of these foods.

YOU NOW KNOW

- The more energy you need, the more Calories you need.
- Weight depends on age, height, body size and work.

4. SPECIAL DIETS

When people speak of "being on a diet," they usually mean a weight-reducing diet or a weight-gaining diet. But there are other reasons for watching one's diet. There are some people who have a disease called DIABETES (dy-a-BEE-teez). Such people are not able to burn up their sugar completely. Unused sugar is given off in their urine. These people should eat fewer carbohydrate foods. In severe cases, they must also take injections of a drug called INSULIN (IN-su-lin). People whose blood needs extra red blood cells are said to have ANEMIA (uh-NEE-me-uh). Their diets should contain foods which are rich in iron. Perhaps you have met people who develop a rash or hives on their skin after they have eaten certain foods, such as strawberries. Such people have ALLERGIES (AL-ler-jeez) to these foods and must avoid eating them.

5. FOOD FADS

A fad is a new, popular way of behaving. Many people follow a new style because it is exciting and sounds right to them. There are many food fads, most of which turn out to be

Fig. 162–1
Exercise is important, but vigorous exercise too soon after meals can injure your health.

quite silly. Have you heard fish called a "brain food"? Actually, fish has no more value for the brain than any other nourishing food has. Every year, there seem to be new fads: "pep" foods, "good skin foods," "tired blood" foods. Food fads are usually worth a good laugh but not the money you are sometimes asked to spend. At the same time, however, there are many real improvements being made in dieting. Some foods, like milk and bread, are *enriched*. Carefully measured amounts of vitamins are added to them. This is done under government control. Sometimes doctors have their patients take vitamin pills. However, many people spend a great deal of money each year buying needless vitamin pills. This is one reason why it is important to know the difference between food fads and real improvements in dieting. Remember: For most people, a balanced diet will keep them in good health.

YOU NOW KNOW

- Diet control depends on a knowledge of science.
- If you need to diet, you should get a doctor's advice.
- Good health depends on good eating habits and knowledge.

6. FOOD HABITS

Much of what you eat, when you eat, and how you eat is a matter of habit. You get used to doing things

Fig. 163–1
A clean school cafeteria helps pupils enjoy their food and digest it properly.

a certain way. Therefore, all people, especially teen-agers, should develop good eating habits. It is usually a good habit to eat at regular times. Eating candy and chewing gum between meals may upset your digestion and ruin your appetite. Some people eat too fast and do not chew their food completely. This also interferes with good digestion. It is a good idea to avoid exercise right after a meal. One of the worst habits is to eat only a few kinds of food. For the best health, you should eat a variety of foods.

7. OTHER PEOPLE'S FOOD

Have you ever heard the expression "One man's food is another man's poison"? This means that we all have different tastes. For example, different nations have developed different eating habits and various kinds of foods. As we have already learned, there are a few basic food requirements needed to carry on the life functions. But there are many foods and many ways of preparing food which meet these basic requirements. The diets of other people may be very different from our own and yet fulfill the same needs. Intelligent people respect other people's tastes. Actually, many of us have learned to like foods of other people. It is an adventure to try new foods. Americans now eat many foods which were once Italian, French, Jewish, Russian, and Irish. Can you name a few such "foreign" foods?

8. SUGAR AND SPICE . . . AND EVERYTHING NICE

From ancient times, man has *added* small amounts of substances to his food. Why? Originally, it was to stop the food from spoiling. Salt, sugar, pepper and mustard were known to *preserve* foods. A material added to a food is called a FOOD ADDITIVE (AD-i-tiv). Man soon found that some substances also gave the food a better flavor or *spice*, and made it "very nice."

Today, food manufacturers use many food additives. But there are now other uses in addition to preserving or improving flavor. For example, we now enrich foods by adding vitamins and minerals. Some chemicals spread the particles around evenly or thicken the food. For example, ice cream and salad dressings need this kind of treatment. Some chemicals bleach the food to make an attractive color, as in drinks, cheeses and jellies. Some are used to keep moisture in the food.

Are additives dangerous? Some may be, even in small amounts. Some are dangerous when over-used. Some additives might fool the consumer by making the food look fresher or more attractive than it really is. The U.S. Food and Drug Administration (FDA) "keeps an eye" on food additives to protect the consumer. FDA inspects foods, checks package labels, and prohibits the use of some additives.

UNDERSTANDING WHAT YOU HAVE READ

I. What are the two main ideas of Chapter 26?

1. Dieting should be done for a definite purpose.
2. Overweight is dangerous.
3. A balanced diet uses all groups of foods.
4. All people have some poor food habits.

II. Explanation please . . . answer in one or two sentences.

1. What kind of foods help keep blood healthy?
2. Which foods may cause overweight?
3. Why may 150 pounds be the right weight for one 44-year-old man but wrong for another 44-year-old man?
4. What is a scientifically chosen diet?

III. Knowing What and Why

1. The Basic Four Groups of Foods are used to
   a. choose diets for people with allergies.
   b. find a balanced diet.
   c. avoid anemia.
   d. guard against food fads.
2. If a person is very overweight, he should immediately
   a. stop eating so much.
   b. follow the Basic Four.
   c. suspect diabetes.
   d. consult a doctor.
3. A drug that is useful in treating diabetes is
   a. insulin.
   b. phosphorus.
   c. iodized salt.
   d. alcohol.
4. Enriched white bread has extra
   a. proteins.
   b. Calories.
   c. vitamins.
   d. calcium.

IV. Match the words in Column A with those in Column B.

A	B
1. anemia	a. the addition of vitamins
2. Calorie	b. food fad
3. brain food	c. a bad reaction to strawberries
4. allergy	d. lack of iron
5. enriched	e. measurement of heat
	f. calcium

V. Unscramble the letters.

The first letter of each word spells a *feeling*.

1. ROPTIEN (a nutrient)
2. LALEYGR (a bad reaction)
3. ILSNIUN (a drug)
4. TINRONEG (an element)

Now write the first letter of each word in your notebook and you will get a word that describes a feeling.

__ __ __ __

Glossary: What Does It Mean?

additive	AD i tiv	A chemical added to food to preserve it, flavor it or otherwise make it more attractive.
alcohol	AL ko hol	A drug found in wine, liquor and beer which produces drunkenness.
allergies	AL ler jeez	Bad reactions to certain foods or surroundings.
anemia	uh NEE me uh	A condition in which blood does not have enough red blood cells.
aquaculture	AK wuh kul chur	An experimental method for "farming" the sea for food.
ATP	A-T-P	A complicated compound in all cells which stores and releases energy.
balanced diet		A diet which supplies all the nutrients in the right amounts.
Benedict's solution	BEN a dikts	A chemical used to test for the presence of sugar.
beverage	BEV er ij	A drink, such as coffee, tea, soda.
Biuret solution	by a RET	Chemical solution used to test for the presence of protein.
Calories	KAL o reez	Units of heat energy in foods.
carbohydrates	kar bo HY drayts	One group of nutrients formed from compounds of carbon, hydrogen, oxygen.
casein	KAY seen	A pure protein.
caviar	KAV yar	Eggs of sturgeon.
Clinitest pills	KLIN i test	Tablets used to test for the presence of sugar.
diabetes	dy a BEE teez	A disease in which an excess of sugar appears in the blood.

DNA	D-N-A	A complicated chemical compound found in the cell's nucleus. It controls heredity.
elodea	el oh DEE uh	A green plant living in water.
enriched food		A food to which vitamins have been added.
filtration	fil TRAY shun	Process of removing undissolved waste from water.
indophenol	IN doh FEE nul	Chemical used to test for vitamin C.
insulin	IN su lin	A drug used to burn excess sugar in persons who have diabetes.
kelp		A large brown seaweed.
mariculture	MAH ri kul chur	Another name for aquaculture.
nutrients	NEW tree ents	Chemical compounds in food used to carry on the life functions.
nutrient tests		Chemical tests for the presence of nutrients.
oxidation	ox i DAY shun	The chemical union of a fuel with oxygen.
photosynthesis	fo to SIN thuh sis	The making of food by green plants.
plankton	PLANK tun	Floating or drifting life in the sea; algae, fungi, protozoa.
polluted	puh LOO ted	Dirty; containing wastes.
poultry	POLE tree	Group of birds eaten for their flesh.
respiration	res per AY shun	Process of oxidizing food to release energy.
rickets	RIK ets	Bone disorder caused by a lack of vitamin D.
roe	ROW	Eggs of fish.
Sargasso Sea	sar GAS oh	An area in the Atlantic Ocean, northeast of the West Indies.
Sargassum	sar GAS um	Large brown seaweeds living in the Sargasso Sea.
scurvy	SKUR vee	Disorder caused by a lack of vitamin C.
seaweed		Algae made of many cells and visible to the eye without a microscope.
Ulva	UL vuh	A green seaweed: "sea lettuce."
vitamins	VY tuh mins	A group of chemical nutrients which keep our bodies working correctly.

Appendix

WHAT IS AN APPENDIX?

An "appendix" is something "added on." The appendix of a book is a section which has special information which can be useful. In this section, you will find all kinds of "odds and ends."

ELEMENTS OFTEN MENTIONED IN BIOLOGY

Name	*Symbol*
Hydrogen	H
Carbon	C
Nitrogen	N
Oxygen	O
Sodium	Na
Phosphorus	P
Calcium	Ca
Iron	Fe
Iodine	I
Radium	Ra

COMMON MEASUREMENTS

English Scales		Metric Scales
12 inches = 1 foot 3 feet (36 inches) = 1 yard 5,280 feet = 1 mile	← distance →	100 centimeters = 1 meter 1000 meters = 1 kilometer
16 ounces = 1 pound 2000 pounds = 1 ton	← weight →	100 grams = hectogram 1000 grams = kilogram
2 pints = 1 quart 4 quarts = 1 gallon	← liquid volume →	1000 ml (milliliters) = 1 liter

ENGLISH AND METRIC SCALES COMPARED

1 inch	=	about 2½ centimeters (the inch is bigger)
1 foot	=	about 30 cm
1 meter	=	about 39 inches (the meter is 3⅓ inches longer than the yard)
1 kilometer	=	⅝ of a mile (the mile is longer)
454 grams	=	1 pound
1 quart	=	95/100 of a liter (the *liter* is bigger)
1 gallon	=	3¾ liters

FORMULAS OFTEN MENTIONED IN BIOLOGY

COMMON NAME	CHEMICAL NAME	FORMULA
Boric acid	Boracic acid	H_3BO_3
Carbon dioxide	Carbon dioxide	CO_2
Carbon monoxide	Carbon monoxide	CO
Ether	Ethyl oxide	$C_4H_{10}O$
Glucose	Glucose	$C_6H_{12}O_6$
Grain alcohol	Ethyl alcohol	C_2H_5OH
Limewater	Calcium hydroxide solution	$Ca(OH)_2$
Peroxide	Hydrogen peroxide	H_2O_2
Salt (table)	Sodium chloride	$NaCl$
Starch	Starch	$C_6H_{10}O_5$
Sugar (table)	Sucrose	$C_{12}H_{22}O_{11}$
Vinegar	Acetic acid	$HC_2H_3O_2$
Wood alcohol	Methyl alcohol	CH_3OH

SOME IMPORTANT TEMPERATURES

Degrees Celsius		*Degrees* Fahrenheit
−273	Absolute zero. All motion of molecules stops.	−460
− 79	Carbon dioxide gas becomes solid (dry ice)	−110
− 39	Mercury freezes	−39
0	Ice melts; water freezes	32
20	Average room temperature	68
37	Normal human body temperature	98.6
65	Pasteurization of milk (for 30 minutes)	149
100	Water boils (to steam)	212

U. S. GOVERNMENT AGENCIES CONCERNED WITH BIOLOGY

Department of Interior: Fish and Wildlife Service; Bureau of Sports Fisheries and Wildlife; National Park Service.
Department of Agriculture: Food and Nutrition Service; Forest Service; Soil Conservation Service; Agricultural Research Service.
Department of Health, Education and Welfare: Public Health Service; Food and Drug Administration; National Institute of Health.
Department of Commerce: National Oceanic and Atmospheric Administration

MAGAZINES FOR STUDENTS OF BIOLOGY

Audubon Magazine
National Audubon Society
1130 Fifth Ave.,
New York, N.Y. 10028

Current Science
American Education Publications
1250 Fairwood Ave.,
Columbus, Ohio 43216

National Geographic
Washington, D.C. 20036

Natural History
American Museum of Natural History
Central Park West at 79th St.,
New York, N.Y. 10024

Popular Science Monthly
Popular Science Publishing Company
355 Lexington Ave.,
New York, N.Y. 10017

Science and Math Weekly
American Education Publications
1250 Fairwood Ave.,
Columbus, Ohio 43216

Science Digest
959 Eighth Ave.,
New York, N.Y. 10019

Science News Letter
1719 N. St., N.W.,
Washington, D.C. 20006

Science World
Scholastic Magazines, Inc.
50 West 44th St.,
New York, N.Y. 10036

The Smithsonian
Smithsonian Institution
Jefferson Drive
Washington, D.C. 20560

TRIPS FOR STUDYING LIFE SCIENCE

There are many communities which have places where you can study life science. The list below is NOT complete. Trips to such places can be fun. REMEMBER THE MOTTO: *Enjoy, do not destroy!*

ZOOS

Bronx, N.Y.	N.Y. Zoological Park
Chicago, Ill.	Lincoln Park
Detroit, Mich.	Belle Island
New York, N.Y.	Central Park Zoo
Philadelphia, Pa.	The Zoo
Pittsburgh, Pa.	The Highland Park
San Francisco, Calif.	Golden Gate Park
Washington, D.C.	National Zoological Park

BOTANICAL GARDENS

Boston, Mass.	Arnold Arboretum
Bronx, N.Y.	New York Botanical Gardens
Brooklyn, N.Y.	Brooklyn Botanic Garden
Jamaica, N.Y.	Jamaica Bay Wild Life Refuge
Coconut Grove, Fla.	Fairchild Tropical Garden
Lisle, Ill.	Morton Arboretum
Ottawa, Canada	Dominion Arboretum
Philadelphia, Pa.	Bartram's Garden
St. Louis, Mo.	Missouri Botanical Garden

San Francisco, Calif.	Golden Gate Park
San Marino, Calif.	Huntington Botanical Gardens
Seattle, Wash.	University of Washington Arboretum

MUSEUMS

Boston, Mass.	Museum of Science
Buffalo, N.Y.	Museum of Science
Cambridge, Mass.	Museum of Comparative Zoology
Cleveland, Ohio	Cleveland Health Museum
Cleveland, Ohio	Cleveland Museum of Natural History
Chicago, Ill.	Chicago Natural History Museum
Espanola, N. Mex.	Ghost Ranch Museum
Los Angeles, Calif.	Los Angeles County Museum
New York, N.Y.	American Museum of Natural History
Oak Ridge, Tenn.	American Museum of Atomic Energy
Ottawa, Canada	National Museum, Natural History
Pittsburgh, Pa.	Carnegie Museum
Philadelphia, Pa.	Academy of the Natural Sciences
Washington, D.C.	Smithsonian Institution

AQUARIA

Fort Myers, Fla.	Florida Marine Museum
St. Augustine, Fla.	Marineland
Woods Hole, Mass.	Fisheries Aquarium
Brooklyn, New York	New York Aquarium

Acknowledgments

The photographs included in this text, on the pages indicated below, appear courtesy of the following:

American Heart Association, 57
American Museum of Natural History, 6, 11, 19, 68, 71, 104, 151; B. Hunter, C. Bogart, 7 (both); J. E. Thompson, 69 (bottom)
American Red Cross of Greater New York, Nate Silverstein, 110
Bausch and Lomb, 83, 87
Central Scientific Supply Company, 83
Center for Media Development, Inc., 30, 128, 155, 157
Eastern Airlines, 41
Goldfeld, Burton, 18
Greater New York Council–Boy Scouts of America, 133
Hamilton Watch Company, 82
Miller, Peter, 152 (right)
Mozer, Lida, 151
Mulvehill, Larry, 9, 13, 31, 62, 135, 162, 163
National Aeronautics and Space Administration, 19
National Dairy Council, 161
National Film Board of Canada, 57
New York City Board of Education, 61
New York City Department of Water Supply, 138
The New York Times, 50
Occidental Petroleum Corp., 152 (left)
Chas. Pfizer Inc., 28
Robinson, Ralph, 15
Secondary School Project–Princeton University, 13
Standard Oil Company, 135
Taylor Instrument Company, 30
Teicher, Samuel, 6, 29, 64
Town of Hempstead, New York, 137
Union Pacific Railroad, 66
United Nations, 32
Weitekamp, Joseph, 69 (top)
Welch Scientific Company, 16, 23

Index